Praise for

PLAY ON

"When it comes to ⬚⬚⬚⬚⬚⬚⬚⬚⬚⬚⬚⬚⬚⬚⬚⬚⬚ in the coal mine — they feel its e⬚⬚⬚⬚⬚⬚⬚⬚⬚⬚⬚⬚⬚⬚ us. In *Play On*, Jeff Bercovici give⬚⬚⬚⬚⬚⬚⬚⬚⬚⬚⬚⬚ler elite athletes use science, strat⬚⬚⬚⬚⬚⬚⬚⬚⬚⬚ keep up with (and often beat) the youngsters. Playing keeps us young, and this is a must-read for anyone who believes age is no reason to quit."

— Bill Gifford, *New York Times* best-selling author of
Spring Chicken and *Ledyard*

"As an athlete who is competing past my so-called prime, *Play On* goes to the heart of some of the biggest questions around longevity and performance that I've been pondering these past few years. It's an utterly fascinating and entertaining blend of science and storytelling that anyone interested in staying fit as they age should read. (And, if you're like me, underline obsessively.)"

— Shalane Flanagan, four-time Olympian,
winner of the 2017 New York City Marathon,
and *New York Times* best-selling author of *Run Fast. Eat Slow.*

"From the surprising science of why Olympic cyclists try to think like kids in competition, to the strategy behind Carli Lloyd's rise from benchwarmer to soccer phenom, *Play On* illuminates a dimension of high-performance sports seldom seen on prime time. Told through an engaging cast of characters, Bercovici's book is a must-read not only for the athlete trying to gain an edge, but for the rest of us interested in living longer, healthier lives. In richly engaging prose, the mysteries of the long and strong careers of today's sports stars are revealed — along with fascinating lessons that could change how all of us view health and fitness."

— Mary Pilon, best-selling author of
The Kevin Show and *The Monopolists*

PLAY ON

*The New Science
of Elite Performance
at Any Age*

JEFF BERCOVICI

*Mariner Books
Houghton Mifflin Harcourt*
BOSTON   NEW YORK

For information about permission to reproduce selections from
this book, write to trade.permissions@hmhco.com or to
Permissions, Houghton Mifflin Harcourt Publishing Company,
3 Park Avenue, 19th Floor, New York, New York 10016.

hmhco.com

Library of Congress Cataloging-in-Publication Data
Names: Bercovici, Jeff, author.
Title: Play on : the new science of elite performance at any age /
Jeff Bercovici.
Description: Boston : Houghton Mifflin Harcourt, 2018. |
Includes bibliographical references and index.
Identifiers: LCCN 2017061532 (print) | LCCN 2017050323 (ebook) |
ISBN 9780544935327 (ebook) | ISBN 9780544809987 (hardback)
ISBN 9781328595966 (paperback)
Subjects: LCSH: Physical fitness for older people | Sports for older people.
| Exercise for older people. | Aging — Physiological aspects. | Sports sciences. |
BISAC: SCIENCE / Life Sciences / Anatomy & Physiology
(see also Life Sciences / Human Anatomy & Physiology).
Classification: LCC GV482.6 (print) | LCC GV482.6 .B47 2018 (ebook) |
DDC 613.7/0446 — dc23
LC record available at https://lccn.loc.gov/2017061532

Book design by Kelly Dubeau Smydra

Printed in the United States of America
DOC 10 9 8 7 6 5 4 3 2 1

For Robyn

You were always the jock in the family

CONTENTS

INTRODUCTION

Overtime Starts Now

'm halfway through the most grueling four minutes of my life, wondering how I'm going to survive the next 120 seconds, when my tormentor, a deceptively personable southerner named Joel, breaks into my suffering with what I assume to be a stupid joke. My assumption is wrong, but it will be another two minutes before I realize that.

"Not to psych you out or anything," Joel says, "but a bunch of coyotes are watching you."

"Hilarious," I manage to croak, and go back to watching the numbers on the digital readout in front of my face crawl upward.

Why shouldn't there be coyotes watching me, though? They're a common-enough sight, prowling for rabbits and the occasional unwary house cat, here on the dusty outskirts of Phoenix, Arizona. I can imagine what they would see: a 38-year-old man, shortish of stature and slight of build, hanging by all four limbs from a black steel apparatus, grimacing and wheezing like he's on the brink of multisystem organ failure. An easy if stringy meal for a pack of scavengers, once they peel off the sweaty sausage-casing of compression apparel.

Granted, this particular spot is an unlikely venue for a wild-animal attack. I'm in a 31,000-square-foot, $10 million state-of-the-art training and wellness facility operated by a company called Exos. Formerly known as Athletes' Performance, Exos owns a chain of these clinics, scattered mostly across the Sun Belt, in cities like Los Angeles, Atlanta, San Diego, Dallas, and Pensacola. Its locations correspond roughly to the regions where professional athletes spend their increasingly brief off-seasons. This facility is the flagship.

Although its clients include large corporations and the U.S. military, Exos is best known for getting some of the world's strongest and fastest people in peak shape before their competitive seasons begin. On this July day, the campus is hosting a contingent of 30 or so NFL players, here to put a scalpel edge on their fitness before reporting to their teams' respective training camps in three weeks' time. This being Phoenix, they had to get their outdoor work out of the way early, before the temperature broke triple digits. At seven this morning, notebook in hand, I watched from a minimum safe distance as LeSean McCoy of the Buffalo Bills, Cameron Jordan of the New Orleans Saints, Prince Amukamara of the New York Giants, Colin Kaepernick of the San Francisco 49ers, and other genetically blessed multimillionaires performed football-specific fitness drills — sprinting between different-colored cones while Brett Bartholomew, Exos's lead NFL trainer, called out directions; accelerating laterally from a standstill while being dragged in the other direction by bungee cords; and so on. The players were clearly working hard, but it was a different sort of hard work than the kind you see dramatized in those Gatorade and Nike commercials. There were no faces contorted in agony or voices screaming out "Give me one more!" The emphasis was on detail, not all-out exertion. Afterward, Bartholomew explained that getting players ready for a season doesn't have much to do with "the stuff they obsess about at the combine" — 40-yard-dash times, bench-press repetitions, that kind of thing. "With actual veterans, it's about taking care of their bodies and doing everything the right way," he says.

Now it's midday and the torrid desert sun is directly overhead: 108°F and 9 percent humidity. The players were sent home hours ago with orders from Bartholomew to spend the afternoon taking a nap, lying in a whirlpool, doing yoga, playing video games — anything but exercising more. "Remember, our muscles have currency just like our bank accounts," he admonished them before dismissing them to the cafeteria to pick up their personalized recovery smoothies, each one tailored to its drinker's body composition and sweat chemistry. (To make sure the athletes are drinking enough fluids, the bathrooms are decorated with posters, conveniently hung over the urinals, showing what pee looks like at different levels of hydration.) It seems like everyone has listened, except for Kaepernick, who proceeds to spend the next four hours in the facility's airplane-hangar-sized weight room, chiseling new creases into his already famously buff, tattooed torso.

With nothing better to do, I sign up for a training session with Joel Sanders, a veteran strength and conditioning coach. A rangy, earnest 30-year-old from Savannah, Sanders focuses mostly on Exos's civilian clientele, like the Silicon Valley venture capitalist who, once a quarter, brings his favorite portfolio companies' CEOs here for a long weekend of boot camp–style workouts, hot and cold pools, organic meals, and deep-tissue massages. Joining me for the session is Dan, a tall drink of water who turns out to be a recently graduated college basketball player staying in shape in hopes of getting a call from an injury-stricken NBA team in preseason.

At Joel's instruction, we start by loosening up on foam rollers, seesawing back and forth over them to knead out our legs and backs. I take this opportunity to ask Joel the question on my mind: What's the difference between me and Dan? Besides, obviously, the 10 inches in height and all the other inborn athletic traits that made him an NBA hopeful and me a natural-born journalist. No, what I'm wondering about is the significance of the 16 years that separates us in age. What's the difference, athletically speaking, between someone at 22 and that same person at 32 or 42 or 52?

"Put your hand on your chest," Joel instructs me. I do so. "Now, take your middle finger and tap your chest as hard as you can." I do that, too: *tap, tap, tap.* "Keep your hand there." Joel reaches out, grabs my middle finger, pulls it back, and lets it go. Without me doing anything, it snaps back against my chest: *THWAP!*

The difference between *tap* and *THWAP!*, he says, is power. In physics terms, power is units of work divided by units of time. In layman's terms, it's the ability to deliver force quickly. If you can lift a 100-pound dumbbell and carry it across the gym, you're strong. If you can pick it up and hurl it across the gym, you're strong *and* powerful. When you admire the explosive first step of a point guard, the 140-mile-per-hour serve of a tennis player, a running back's burst through the hole, a volleyball hitter's thunderous spike, what you're admiring is power. More than anything else, it's power, Joel says, that separates young athletes from old ones. It's what we're here to work on.

We begin with a dynamic warm-up: Elastic bands around our knees, we shuffle forward and backward, side to side. We run in place on our toes, then drop into sudden squats, activating our quads and glutes. Moving like inchworms, we walk our hands out in front of us, then walk our feet up to them until we're bent double. Sweating lightly now, we start the workout proper. Joel gives us each a rubber medicine ball — Dan's big, mine little — and has us jump and hurl it toward the faraway ceiling a dozen times, then twist and slam it on the floor a dozen more. We do shoulder presses, cable rows, single-leg squats, and Romanian deadlifts. It's a good workout, hard but fun. I manage not to embarrass myself in front of Colin Kaepernick, who is still over in the far corner of the gym, doing something complicated with kettlebells. I think we're done.

One more exercise, Joel says. He leads me over to a machine standing by the wall of windows. This, Joel says, is the VersaClimber. Every athlete who works out at Exos hates it. In fact, I will learn later, athletes everywhere loathe it. When the Cleveland Cavaliers' strength coach needed to get his team ready for the NBA's endless playoff schedule during its championship 2016 postseason run, it was the VersaClimber he turned to. Forward Tristan Thompson said bonding over their shared hatred of the VersaClimber helped cement the team's chemistry. Mark Sisson, an elite Ironman competitor and fitness guru, calls the VersaClimber "the greatest piece of fitness equipment ever invented," one that would be more popular except it's "too intense an experience for most people. The few that have used it inevitably quit because it's so hard." The way it works is simple as can be: you put your feet on the pedals, grip the handles, and make a climbing motion. A digital readout shows how far you've ascended in place. My goal, which Joel tells me should be doable for someone of my fitness characteristics, is to climb 400 feet in four minutes.

Four minutes. How hard could that be? (At age 61, Sisson was doing four intervals of 1,000 feet each.) I start. Within 30 seconds, I realize: four minutes could actually be extremely hard. Within a minute, I think, *I don't know if I can do three more minutes of this.* After another 30 seconds, I'm not thinking anything because all the glycogen in my body is rushing to my muscles to replace my zonked-out stores of adenosine triphosphate (ATP), leaving none left over to power my frontal cortex. Then, at around the two-minute mark, Joel makes his joke-but-not-a-joke about the coyotes. I barely clock it. I can't explain why this particular activity is so much more unbearable than literally any other I've ever done, except that there is no respite for any part of me. When you're cycling up a steep hill, say, your quads and glutes may be begging you to quit, but that's just your legs. On the VersaClimber, there is nowhere to hide, no part of your body in which your consciousness can hole up and wait it out. But it's only four minutes, right? By the time I pass three minutes, I'm pretty sure I'll be able to finish in some form, but it's even odds I will immediately punctuate that accomplishment by throwing up a stomachful of neon-yellow electrolyte beverage. Finally, the clock reaches 4:00.

Total feet climbed: 399.

Joel places a steadying hand on my back as I step off the pedals and stagger drunkenly around the machine. He hands me a sports drink, eases me to the ground, and has me put my legs on a vibrating platform to help with muscle recovery. Just take it easy for a few minutes, he says. My head lolls to the side, and that's when I see Joel wasn't joking after all. They were watching me. Coyotes, half a dozen of them. With a capital *C*. That is, members of the Arizona Coyotes, the local NHL team. Sitting in a semicircle in the stretching area, loosening up with foam rollers, they witnessed my whole ordeal, and they're still eyeing me with curious looks. For my workout to have held the attention of these professional athletes, who see exercise-induced suffering every day, I must have looked even more alarming than I felt. I imagine what they must have been thinking: *Who is this little dude who's about to die in our gym and what is he doing here?*

What *am* I doing here?

Actually, that's pretty easy. I went to Exos on a mission to understand a certain subset of elite athletes: the ones who continue to perform and compete at the very highest levels long after the age when most of their peers have faded away. I want to know what makes them different from everyone else, what secret sauce allows them to defy the conventional wisdom that high-level sport is a young man's (or woman's) game. I want to know how they work out and recover, how they eat and sleep, how they stay healthy and heal from injuries, what's in their DNA and their mitochondria, how they think and learn and strategize and motivate themselves to keep getting better through the seasons and decades. In the answers to those questions, I believe, lies the key to something so many of us desire more than almost anything else: the ability to stay healthy and vital and competitive as we get older; to feel like we possess a measure of control over how we age; to experience the passage of years as an unfolding of new possibilities rather than a long hallway of closing doors.

My quest has led me to training camps, tournaments, research centers, hospitals, antiaging clinics, medical conventions, and technology conferences. I've interviewed Super Bowl champions, Olympic gold medalists, World Cup soccer players, big-wave surfers, and backcountry skiers, geneticists, biomechanics experts, inventors, sports psychologists, orthopedic surgeons, elite Special Forces operators, and people whose professions, like the self-taught health gurus who minister to the bodies of Tom Brady and Serena Williams, defy categorization. My goal was to find out everything they could tell me about the latest advances, and advances still to come, in areas from nutrition to brain science to virtual reality.

Age and sports: you can't separate them. Try talking about one without the other and it will be a short conversation. If you're over 30, you've probably had the experience of watching a sporting event and hearing commentators endlessly discussing whether an athlete some years younger than you is over the hill yet. An average afternoon of ESPN will feature a dozen different euphemisms for "old." Does Johnny Veteran still have enough gas in the tank? Is there any tread left on his tires? Has he lost a step? Does he still have his fastball? Just as the home run sluggers of baseball's juicing era will have asterisks next to their names in the record books, older competitors are constantly reminded of what makes them different. There's practically a law that any magazine profile or TV introduction of a top player over 30 must contain the qualifying phrase "at the advanced age of . . ."

It's not an arbitrary fixation. At their core, sports are about challenging our physical limits — through effort and grace, talent and grit, teamwork and individual brilliance. Age is the final and most stubborn of those limits, if a relative one. "Old" means different things in different sports. Ultra-endurance athletes often don't peak until after 40, while gymnasts rarely last past 22. A 30-year-old NFL star might be pushing retirement or just entering his prime depending on what position he plays. Still, at every point of one's career, the specter looms. A lucky player can avoid getting hurt, traded, or cut, but nobody avoids getting older. Father Time is undefeated, as the saying goes.

Recently, however, it seems as if it's Father Time who might be losing a step. In the last few years, without a ton of fanfare, we've entered a golden age for older athletes. Everywhere you look across the sports world, individuals are competing, and winning, at ages that as recently as a generation ago would have been considered downright geriatric. Take the events of a single year, 2016. In January, 39-year-old Peyton Manning became the oldest quarterback ever to start, and win, a Super Bowl. (The following January would see another 39-year-old, Tom Brady, match his feat.) In tennis, Serena Williams, defending her Wimbledon title, became the oldest woman, at 34, ever to win a Grand Slam tournament. She would break her own record the following year. In basketball, the San Antonio Spurs fielded the oldest team in the NBA, led by 40-year-old Hall of Fame lock Tim Duncan, and proceeded to win 67 of their 82 regular-season games, including going 41-1 at home. In his 25th NHL season, 44-year-old Jaromir Jagr led his team, the Florida Panthers, in scoring, moved up to third place on the all-time goals list, and didn't even think about retiring. At the British Open, in a showdown critics compared to the legendary "Duel in the Sun" between Jack Nicklaus and Tom Watson, 40-year-old Henrik Stenson edged 46-year-old Phil Mickelson on the strength of a 63-stroke final round, tying a major tournament record. At the Rio Summer Olympics, marathoner Meb Keflezighi became, at 41, the oldest American distance runner ever to compete at the games. Cyclist Kristin Armstrong, who turned 43 in Rio, became the first Olympian ever to win gold three times in the same cycling event. (Her first win came at age 35.) Thirty-four-year-old sprinter Justin Gatlin became the oldest man ever to medal in the 100-meter race. Michael Phelps, the most decorated swimmer in history, also became the oldest ever to win gold in an individual event, at age 31. His record fell after a mere three days, however, when fellow American Anthony Ervin, 35, took gold in the men's 50-meter freestyle race.

It's not a matter of a few outliers. In virtually every sport, athletes are sticking around for an extra victory lap or three rather than shuffling off to the metaphorical showers at the first sign of gray. Between 1982 and 2015 the number of NBA players 35 or older jumped from 2 to 32. In the NHL it went from 4 to 50 over the same span; in the NFL it increased from 14 to 38. In tennis, the average age of the men ranked in the top 10 has increased by more than five years since 1992. The average age of U.S. Olympic swimmers has gone up three years for women and four years for men. And all this is happening despite multiple trends that in theory ought to be shortening careers, from earlier specialization in youth sports to greater awareness of long-term concussion risk to longer competitive calendars and more travel. (Thanks in part to the metastasis of the NBA playoff format, before LeBron James turned 31, he had already clocked more playing minutes than Magic Johnson or Larry Bird did in their entire careers. No wonder his hairline is receding.)

By this point, the cynical sports fan has been screaming "Steroids!" for several paragraphs. The suspicion is not misplaced. Without a doubt, sophisticated doping regimens of the kind made infamous by Alex Rodriguez, Lance Armstrong, and Marion Jones have played a role in extending the careers of elite athletes, especially the ones who have shown surprising late-career surges or miraculous bounce-backs from chronic injury. Although it's difficult to quantify the prevalence of illicit activity, doping is likely only becoming more common as the increasing financial rewards of sports fame induce athletes to delay retirement. There's a reason the shady medical practitioner at the center of so many doping scandals inevitably runs what's referred to in the press as an "antiaging clinic." That's less of a euphemism than it might seem; in many ways, as we'll see, the point of performance-enhancing drugs is no more or less than to make older bodies mimic younger ones in the way they heal from damage, recover from training, pack on muscle, and turn oxygen into fuel. But doping, as widespread as it is, is just a small piece of a much bigger and more interesting phenomenon — one that's by no means confined to professional sports.

You're probably familiar with the phenomenon I mean. Since you're holding this book, I'd guess you're part of it. I'm talking about the massive, decades-long shift in the way we — including the 99 percent of us who aren't elite athletes — think about the relationship between age and physical activity. All over the developed world, but especially in the U.S., vast numbers of adults are integrating sports and fitness into their lives in a way that simply wasn't the case a generation or two ago. It's hard to comprehend from a modern vantage point, but 50 years ago, the notion of an average 40-year-old jogging for exercise or going to the gym to lift weights was a novelty. An adult athlete was someone who played golf or softball or perhaps, at the extreme end, tennis. In just a few short decades, participatory sports have come to saturate our culture. In the San Francisco Bay Area, where I live, there's not a day I step outside my house that I don't see people my age and older wearing brightly colored workout apparel and minimalist running shoes and heart-rate-tracking watches, riding $5,000 racing bikes and kite boards, practicing with their Ultimate Frisbee or rugby teams, training for 5Ks or gran fondos or ultramarathons. The numbers bear it out. Participatory sports and fitness is an $85 billion business in the U.S., one that's growing significantly faster than the rest of the economy. Participation in the World Masters Games has tripled since the first one was held in 1985; the 2017 edition drew more than 25,000 over-40 athletes all the way to New Zealand. Marathons get more popular with each passing year, with over-50 runners representing one of the fastest-growing cohorts. Adult recreational leagues in big cities are selling out as team sports supplant golf as a business-networking venue; "Soccer Is the New Golf," declared a 2012 headline in *Crain's* business magazine. This culture shift achieved its apotheosis in Barack Obama, the first post–baby boom president and the first to run a regular pickup basketball game out of the White House.

Pain is another way to measure the trend. Sports injuries among postcollegiate adults have been growing at double-digit rates. Surgeries like anterior cruciate ligament (ACL) reconstructions and meniscus repairs, once considered unwise for patients over 40, are now routine. "The age group you see who considers themselves a sports medicine patient is a lot older than would even walk into a clinic fifteen years ago," says Nirav Pandya, an orthopedic sports surgeon in San Francisco.

In the last 10 years, this gradual sea change has become more like a tsunami. If the boomers popularized the idea of adults as athletic beings through pursuits like jogging and aerobics, the generations after them have completed the revolution, advancing the notion that anyone can be a high-performance athlete on their own terms. It's no coincidence that the fastest-growing fitness sports are also the most intense and challenging. CrossFit, the high-intensity circuit-training chain/religion, has 5,000 locations in the U.S.; in 2014 more than 200,000 people entered qualifiers for the CrossFit Open, the sport's equivalent of the Olympics. Some 1.3 million people have done Tough Mudder, a military-style obstacle-course race that challenges participants to run 12 miles while scaling climbing walls, slithering through mud pits, and hauling logs. About half of those racers are over 30. Triathlon enrollment keeps setting new records; membership in USA Triathlon has swelled more than 400 percent since 1998, to more than 550,000, and the number of triathletes over 50 has more than tripled in a decade. Weekend warriors are now the superstars in a new genre of spectator sports. More than 6 million people watched the 2015 season finale of *American Ninja Warrior*, a TV game show in which amateur athletes navigate obstacle courses designed to challenge their strength, agility, and endurance. After seven seasons and thousands of contestants, two men were finally able to complete the show's ultimate challenge: 36-year-old cameraman Geoff Britten and 33-year-old rock climber Isaac Caldiero. An adaptation of a Japanese television show, *Ninja Warrior*'s popularity has inspired several knockoffs. All across the country, *Ninja* hopefuls train on obstacles they've re-created in their own backyards, hoping for an everyman's shot at glory.

I'm part of this phenomenon, too, in a modest way. As a kid growing up in Wisconsin, I played all the usual sports, and was even decent at one or two. I was never graceful or coordinated, but I was fast as a jackrabbit, the kind of kid coaches kept around to demonstrate "hustle" for the more talented but less motivated. In high school, I self-segregated with the other creative nerds, and that was more or less that for my career as a jock for the next 15 years. I got back into sports in a big way at 33, when a friend invited me to join her co-ed recreational soccer team. It was a casual league, she promised me over beers. I'd just gotten divorced and moved into a depressing efficiency apartment in Brooklyn, and figured I could use some new friends and a hobby that didn't involve a glowing screen. "Are you any good?" my friend asked. "No," I told her, "but I'm fast."

And I was, in that first game, for about three minutes. After that, I was gasping, cramping, waving weakly at the sideline in hopes a substitute would come on and end my humiliation. At one point, an opposing player I was marking cut hard with the ball. I tried to mirror the move, but my legs had other ideas. I buckled to the ground as though gravity had suddenly tripled. Lying there on the damp Astroturf, I understood the difference between 23 and 33 for possibly the first time. (I later found out the "casual league" was suspiciously full of former Division 1 players.)

Naturally, I was hooked. Over the next couple of years, I made it my mission not to be the worst player on the field, even when I was the oldest, which was often enough. The more I played, the better and fitter I got. Having started out in soccer feeling prematurely middle-aged, I now had the wonderful sensation of aging in reverse. *Not bad for a 35-year-old,* I'd tell myself after outracing some recent college grad to a ball in the corner.

Then, the nagging hip and back injuries started to pile up. I got X-rays and MRIs that didn't tell me anything. A doctor friend told me my options were to stop playing or up my Advil dose. Was I too old for this, after all? But as I frequented more pickups and league games around the city, I encountered players in their 50s or even 60s who amazed me with their stamina and skill. What were they doing that I wasn't? I'd always been a sports fan, but I started paying attention to players' ages, rooting for the older ones on reflex. Watching the Olympics one Saturday at my then girlfriend's house, ice packs strapped to both hips like a cowboy's six-shooters, I grew excited when an announcer mentioned that Ryan Giggs, the Manchester United winger and British national teamer, was three years older than I. "Aly! The captain of Great Britain's soccer team is thirty-eight!" I yelled upstairs.

"That's great, babe," came the reply. "You're not going to play in the Olympics." (We're married now, obviously.)

My heightened interest turned into something more like obsession after I suffered two herniated disks in my back during an indoor game, ruptures so severe it required emergency neurosurgery to prevent the fast-spreading numbness and weakness in both legs from progressing into permanent paralysis. The surgery worked, but the nerve damage left me so frail I could barely walk around the block. On one convalescent outing, I remember cringing in fear as a rambunctious child of maybe eight came running in my direction, looking the wrong way. *So this is what it's like to be 90*, I thought when the danger had passed.

It was eight weeks before I could bend sufficiently to put on my own socks and six months of intensive physical therapy before I was cleared to start jogging again. In those first long weeks, with little to do except lie on my couch and wait for the buzzing in my sciatic nerve to subside, I formulated a new game plan: I was going to play soccer again. Not only that, but I was going to get on top of this aging thing. And I was going to do it by finding out how the older athletes I watched on TV did it and copying as much of it as I could. If athletes like Ryan Giggs could compete at the highest level in the world, in the most physically demanding sports, at my age or older, I thought, surely it couldn't be that hard to convince my body to cooperate with my relatively modest demands.

In this, too, I was part of a bigger movement. Since the first Olympiads were held in ancient Greece, professional athletes have always been a source of public fascination. But as we fans have become athletes in our own right, we're watching them in a whole new way. What we crave now is not so much vicarious thrills as inspiration and information. We're less interested in what LeBron's really like than we are in his postgame cryotherapy regimen. We watch Novak Djokovic outlast opponents in epic baseline rallies and wonder if going gluten-free might improve our stamina, too. If we need arthroscopic surgery, we take extra comfort, and maybe a little unearned pride, in knowing our surgeon is the same guy who did A-Rod's hips. Thanks to the explosion of wearable sensors and advanced data-analysis tools, we can measure ourselves against the pros in a way we never could before. Within the next couple of years, virtual and augmented reality will even make it possible to experience what it's like to be them in a visceral way.

There's more than a dash of aspirational thinking in all of this, of course. Working out like Carli Lloyd or eating like Kobe Bryant is not going to make you Carli or Kobe. But that's not really what we're after. We're after answers, and we look to the elite sports world because we know that's where they'll show up first. When it comes to separating real knowledge from junk science and superstition, an environment where hundredths of a second and fractions of a centimeter matter is the ultimate crucible for what works. In an era when individual sports stars can sign billion-dollar sneaker deals and television rights sell for many times that, the body of a world-class athlete is one of the most valuable pieces of hardware on earth. Every second of that athlete's day has a price tag on it; every additional year of health is worth tens of millions. If you're that athlete — or his coach or doctor or physiotherapist or nutritionist or sleep consultant — you're going to make darn sure you know the cost-benefit before you add something to your performance program. (That's not to say elite athletes don't swear by some questionable practices, like sleeping on the bare ground, or bathing in red wine. But even the fringy stuff can have surprising benefits.)

For elite athletes, however, it's not enough just to do what works; they want to be the first to do it, before the competition catches on. A new technology — like a brain-stimulating headset that promotes faster motor-skill acquisition, or stem cell injections that cause worn-down cartilage to regrow — just might be the difference between a championship season and a rebuilding year. For that reason, sports superstars are often willing to act as guinea pigs for innovations that are still being tested and refined. "Five years from now, we'll know it works and it will be well accepted. But in those five years, your career comes and goes," says Stuart Kim, a Stanford University scientist who studies the genetic basis of longevity in athletes. "So if you're in the upper echelon, you have to do some pretty crazy things." If it's new in the realms of fitness, nutrition, movement science, orthopedic or rehabilitative medicine, or high-performance materials, you can be sure top athletes have access to it first, long before consumers get a sniff. One study of the way medical innovations spread found a 17-year lag between the time a new technique is pioneered and the time it filters down into the general population. Peering into the world of elite athletes is literally a glimpse into the future the rest of us will soon be living in.

It's an exciting future. But what's most exciting isn't the gee-whiz gadgetry or the medical miracles just over the horizon. It's the knowledge we can harness today. To be sure, cutting-edge technology and medicine are helping older athletes succeed in a way they never have before. But there's so much more to the story. Thanks to decades of research and practice, sports scientists have a completely new understanding of how to optimize performance at any age, the principles of which anyone can grasp and put to use. It's an understanding that upends our received notions about what older athletes are capable of and what kinds of sacrifices we should and shouldn't make in order to reach our goals. We now know, for instance, that avoiding injuries is not just a matter of luck or fitness but a science all its own, and possibly even a skill you can learn. We know that the volume of training work you do — how many miles you run or circuits you crush — is far less important than how effectively you recover in between trainings. Forget what you see in those Gatorade commercials: the most successful athletes, it turns out, aren't the ones who work out the hardest but the ones who avoid the insidious buildup of fatigue. They're the ones who recognize the difference between beneficial and harmful training stress and adopt methods that maximize the former and minimize the latter. We used to think having a champion's discipline meant being the last one to leave the gym; now we know it more often means being the first (but heading straight to the cold tub, or a massage therapist's table, or off to bed). "Quality, not quantity" and "Work smarter, not harder" have replaced "No pain, no gain" as the motivational slogans on the walls of pro sports weight rooms.

For a long time, we thought victory in sports was a matter of "Faster, higher, stronger." Now we understand that, so much more often, it comes down to keeping the body healthy enough to unleash the more important advantages of the mind: experience, discipline, tactical nous, emotional stability. If that sounds like a distinction that favors the mature over the young, that's no coincidence. Equipped with advanced analytics, touchscreen playbooks, and digital film libraries, today's athletes compete in games that are orders of magnitude more complex than those of yesteryear. If it seems like Tom Brady keeps getting better as he gets older, it's because it now takes 15 years to really master the art of being an NFL quarterback, and the same thing is happening across other sports.

In short, as sports science races into the future, it sends back dispatch after dispatch full of welcome news for today's older athletes. No, there's no fountain of youth or miracle pill that reverses aging, although Silicon Valley is full of billionaires who are prepared to spend millions searching for just that. (In the meantime, study after study points to exercise itself as the unbeatable antiaging remedy.) Even for the greatest physical specimens on earth, getting older means getting a little slower, a little weaker. But finding peak performance as an older athlete isn't about denying reality. Nor is it about reconciling oneself to suffering or disappointment. It's about being the best version of yourself, performing better while feeling better, and for that we don't need to level the playing field, just tilt it a bit.

PLAY ON

1

RACING AGAINST TIME

The Physiology of the
Aging Athlete

I'm still about an hour's drive from my hotel in Mammoth Lakes when my car starts making a sound I've never heard before. It's a sort of keening wail, which builds to a shriek every time the grade turns uphill. The engine feels sluggish to respond when I step on the gas, too. This would be a lousy time for a breakdown. It's well after dark, about 10:00 p.m., and I'm scheduled to meet up with one of America's most accomplished Olympic athletes in just a few hours. Plus I'm deep in the interior of Yosemite National Park, as wild a place as there is in California. My brush with Coyotes may have ended with no worse than light damage to my dignity, but I'm pretty sure the bears in Yosemite are the nonhuman kind. I'm not getting much in the way of cell reception, either, surrounded as I am on all sides by towering peaks. A sign tells me I'm approaching the highest stretch of road in the entire Sierra Nevada range, Tioga Pass, at 9,943 feet.

Then it hits me: it's the altitude, of course. There's nothing wrong with my little Subaru; it just has to work a lot harder than

usual to burn gasoline up here two miles in the sky, where a cubic meter of air contains 30 percent fewer oxygen molecules than
it would at sea level. You'd think I would've been quicker to grasp
this, seeing as the thin air is exactly what brought me here.

The following morning, as the sun is just sliding up over the
mountains that rim the town of Mammoth Lakes, I park my now-
exonerated vehicle in front of a nondescript brown townhouse situated on a golf course. Into the morning light emerges the elfin
form of Mebrahtom Keflezighi, the preeminent male American
marathoner of his generation. He's ready for some road work.

In the U.S., where distance running is more a phenomenon of
participation than spectatorship, Keflezighi is the rare individual
whose fame transcends the sport. (One measure of his renown:
like Madonna and Oprah, he is referred to by all, or at least by everyone who knows who he is, by a single name — in fact, a single
syllable. The Meb brand is so strong, it feels weird to call him anything else, so that's what I'll call him here.) Like most of the world's
dominant marathoners, Meb was born in the Horn of Africa, in
the tiny nation of Eritrea. Except then, it wasn't yet a nation. Meb
spent his early childhood in a region torn apart by its war for inde-
pendence from Ethiopia; his father, a separatist, lived in constant
fear for his life. When Meb was 11, the family emigrated as refu-
gees, first to Italy, then to San Diego, where his running talent was
discovered when he ran a 5:20 mile in his seventh-grade PE class.

After starring on UCLA's track team, Meb ascended to the world
stage in 2004 with a silver medal in the Athens Olympics, breaking
a 20-year marathon medal drought for Team USA. A stress fracture in his pelvis kept him out of the 2008 games in Beijing, but he
rebounded to win the New York City Marathon the following year
with a then personal-best time of two hours, nine minutes, and 15
seconds. Despite Meb's beating that time the following November,
his sponsor, Nike, apparently felt the then 35-year-old runner was

a depreciating asset and declined to renew his endorsement deal when it expired in 2011. Perhaps the company thought an ultrapolite, soft-spoken, devout Christian would be the sort to take such a slight meekly. But Meb's five-foot-five, 125-pound frame conceals the fiery heart of a ruthless competitor, one who now had a new mission: to make Nike regret its misjudgment. (He also had a new sponsor, Skechers, whose marketing department had a better sense of market timing.) A race-day mishap limited Meb to a fourth-place finish in the 2012 London games, but his greatest moment was still ahead of him. That would come in April 2014, on the first anniversary of the Boston Marathon bombing that killed three and maimed dozens. In an emotionally charged atmosphere, days before his 39th birthday, Meb ran the race of a lifetime, outfoxing a raft of younger and more heavily favored competitors to win the race for his adoptive country.

It was the climax of his career but by no means the denouement. Meb would go on to qualify for the 2016 Olympics in Rio, where, on a rainy, sloppy course, he settled for a 33rd-place finish, out of 155 starters. He slipped and fell on the slick asphalt, landing with his hands on the finish line, and delighted the crowd by breaking off a few push-ups before getting up, sparking a brief "#mebpushups" internet meme. But that's still ahead on this clear October morning in Mammoth. The more immediate goal is the 2015 New York City Marathon, which is three weeks off. Because the Olympic trials are coming up in January, this year's race has less than its usual significance. Marathoners train on long time cycles and most run only a few full races per year. Many of the top American runners have opted to skip New York entirely, lest the long recovery period needed afterward interfere with their training schedules. Meb didn't want to do that — he loves the big-city atmosphere in New York and knows he has only so many more opportunities to enjoy it as an elite — but he's approaching this race more as a tune-up for the trials than the main event. (He'll end up finishing seventh and beating the previous record time for a 40-plus runner by a handy 20 seconds.)

Riding herd on Meb's workout this morning is his coach, Bob Larsen. The two have been together since Larsen, as the longtime maestro of UCLA's cross-country program, coached Meb to multiple NCAA titles in the late 1990s. In the running world, Larsen is a legend in his own right, a member of the Track & Field Hall of Fame and the subject of a documentary feature, *City Slickers Can't Stay with Me: The Coach Bob Larsen Story*. After more than 20 years together, their relationship dynamic is more paternal/filial than professional. Meb bridles when "Coach Larsen" suggests he still needs to lose a couple of pounds before race day, and makes it known that the older man is there to offer advice, not give orders. "He's a mentor but I do decisions for myself," he tells me. "He's not my coach-coach."

Meb, in a red T-shirt and black visor cap, sets off down a gently undulating dirt road at a jog to get warmed up. Larsen and I climb into a white SUV and ride behind him, watching the metronomic two-stroke of his white knee socks. "That looks easy, but he's running a six-forty-five, six-thirty pace," says the 76-year-old Larsen. "I can't run that fast anymore. I'm more like a twelve-minute miler. Ten if I feel really good." The meat of today's training session, he explains, will be what's known as a tempo run. In layman's terms, it's a run that's performed at the runner's maximum sustained pace, the fastest he or she can go for 30 or 60 minutes at a time without overdoing it. A tempo run is often called an anaerobic-threshold run because of the specific physiological adaptation it promotes. Muscle fibers can use two different molecular processes to convert food into energy. Below a certain intensity level, the muscles primarily use oxygen to burn carbohydrates, a form of energy production that can be sustained almost indefinitely. (For ultramarathoners, who compete over distances of 50 miles or more, the need to sleep is often what ultimately forces a halt.) The rate at which they're able to pull oxygen out of the red blood cells carrying it from the lungs limits how fast energy can be produced this way, however. To get above this ceiling, the muscles can tap into another molecular pathway that allows them to burn carbohydrates without oxygen. This anaerobic metabolism, as it's called, has its own downside: it produces lactic acid, which causes muscles to fatigue rapidly as it builds up. This side effect of anaerobic exercise is why even the fittest athletes can only sprint all-out for 10 or 20 seconds at a time. Moreover, the built-up lactic acid requires extra oxygen in order for muscle cells to be able to convert it back into more useful fuels. Shifting into anaerobic mode thus results in an oxygen debt that temporarily lowers the body's aerobic capacity until it's paid off.

Marathon racing is all about the interplay between aerobic and anaerobic metabolism. A runner who gambles and exceeds his aerobic threshold to take an early lead may find himself too depleted to match a rival's surge later in the race. But the runner who falls too far behind might find herself staring at a gap that can't be closed without an unsustainably long detour into anaerobic territory. Timing is everything, and runners are constantly trying to bait one another into showing their metabolic cards too early or holding on to them until too late.

That's how Meb pulled off his signature victory, the 2014 Boston win. Going into the race, he had twice run the course in under two hours, 10 minutes, but in both races he had been hampered by small injuries. This time he was perfectly healthy. On the other hand, he was facing a stronger field than he ever had before: 15 other runners lined up that morning had posted better times than his best. "On paper, he was a two-oh-nine runner versus guys who are that many minutes faster than he is. I don't know how many guys who went to the starting line had run under two-oh-five," Larsen says. "But he makes great decisions in races and that's why he's been successful in winning or placing high in many marathons where people weren't thinking he was going to be all that competitive."

One decision Meb had to make right away was how aggressive to be in positioning himself against the other contenders. Some runners have a particular strategy they want to pursue in every race. Meb is more opportunistic. "He's comfortable leading, he's comfortable hanging on, he's comfortable coming from behind. Very few athletes can do all three of those things, tactically," Larsen says.

The Boston Marathon course can be punishing for frontrunners. It runs in more or less one direction the whole time, and often there is a headwind blowing the other way. Meb woke up that morning planning to "draft" behind other runners much of the race, using them as human windbreaks and conserving energy for a late surge. But after Meb completed his gym warm-up, Larsen rechecked the weather forecast and saw it was looking calm. That meant it should be possible for Meb to run at his own preferred, steady pace, even if it meant getting ahead of the pack — which he proceeded to do 45 minutes in. To a man, the favorites let him go, possibly thinking his 23rd-place finish in the New York City Marathon six months prior meant he was — as Nike had concluded three years earlier — not a serious threat.

The network-television crew covering the race made the same mistake. Even as Meb extended his lead to over a minute, the TV director kept the shot on the group where defending champion Lelisa Desisa and other touted Ethiopians and Kenyans ran together. Riding a shuttle bus from the starting line to the finish on Boylston Street, Larsen tried to keep up with the action by phone, but information was hard to come by. "I'm getting telephone calls saying, 'What happened to Meb? Did he drop out? Because he's not with the lead group,'" Larsen recalls. "And I said, 'I think he might be up in front of them.' Because I didn't think he would have fallen back with the pace they were running."

All this time he's been telling this story, Larsen has been driving. After following Meb along bumpy dirt roads for three miles, we veer off so Larsen can pick up a coffee and a newspaper. ("Up here, if you don't get the paper early on a Saturday, all the *LA Times* are gone," he says.) We meet back up with Meb on the road outside a green church opposite Mammoth's small airport. He glides into the parking lot, sweating freely, having completed 10 miles, gradually ramping up his pace to six-minute miles. The next portion of the workout will be on asphalt, so Larsen hands Meb his lighter road shoes, plus a water bottle. Meb strips off his sweat-soaked red T-shirt and hands it to Larsen, who automatically tosses it to me before remembering I'm not a member of the pit crew. ("Whoops," he says with a shrug.) With his shirt off, you can see why Meb finds Larsen's comment about his weight nitpicky. If there is any spare fat on his frame, you would need a tweezers to pinch it. Although tiny, even by the standards of a sport that favors lighter frames, he's surprisingly well muscled, with a hint of bantam's vanity about it.

For the next 8 miles, we'll be leapfrogging Meb in the truck, getting out every mile to clock his split times as he passes us. This is the tempo run. The goal is to settle into something approximating race pace and hold it for as many miles as feels comfortable but challenging, probably 8 to 10. Meb puts in his earbuds, hits play on some Michael Jackson, and he's off flying. Larsen and I have only a couple of minutes to drive each mile, stop the truck, and locate the faded mile marker painted onto the road shoulder. By the time we look up, Meb is upon us. "About five-oh-five!" Larsen calls out the window to his protégé as we zoom past.

In between timings, Larsen resumes his tale of the Boston Marathon. Meb was in front — way out in front, as some of those sub-2:05 runners finally figured out. With less than four miles to go, two elite Kenyans, 28-year-old Wilson Chebet and 29-year-old Frankline Chepkwony, kicked it into high gear and gave chase. Shaving more than 20 seconds off their mile times, they closed the gap to 30 seconds, then 20 seconds, then 10. With a little over a mile left, Chebet pulled to within 6.2 seconds behind. The previous October, at the Amsterdam Marathon, he had set a course record of 2:05:35, more than three minutes faster than Meb's lifetime personal best. Most observers did the obvious math — a considerably younger, objectively faster guy, comfortably within striking distance — and assumed Meb was roadkill. "When you're within six seconds, they've got you. That's like from here to the truck over there," Larsen says as we hunt for another mile marker. But watching on TV — by now even the network had figured it out — Larsen saw something different. "I said, 'I think Meb's going to hold them off,'" he recalls.

Once again, he was right. As he closed in on the finish, Meb glanced back over his shoulder and saw the race was his. Chebet was no longer closing but losing ground. Feeding off the delirious crowd, Meb indulged in some celebratory fist-pumps, then crossed himself just before breasting the tape. Moments later, he was sobbing. "When the tape touched my chest, all I could do was look up at the sky and say, 'Thank you, God, for giving me this opportunity,'" he says.

How did Larsen know Meb was going to hang on? It was all in the body language. To cut Meb's lead, Chebet had been forced to cross his anaerobic threshold and stay there for way too long. Now the oxygen debt was making itself apparent in his ragged mechanics. Meb, meanwhile, was continuing to run with near-perfect form, compact and linear. Above all was the way his feet were hitting the pavement. Pushing off the ground is how runners produce forward movement, so they need to spend some time on it. (Plus there's that whole gravity thing.) But since the ground isn't moving with them, any time a runner spends in contact with it he is effectively braking. The less ground-contact time, the more efficient the stride. An elite distance runner like Meb typically spends less than 200 milliseconds on the ground with each step.

"Anyone else not used to him would say, 'Oh, he looks great all the time,' but I can see how quickly he's coming off that foot," Larsen says. "Like this." To demonstrate, he pats his hands together, crisply and rhythmically: *pop, pop, pop, pop.* The sound reminds me of Joel's chest-tapping demonstration of power. Meb won Boston, in short, because he was able to finish with power and his opponents weren't. In a real sense, he was operating with the biomechanics of a young athlete, while his vanquished opponents were moving more like older ones.

As I'm making this connection, Meb comes flying by again, completing his seventh mile at tempo. Larsen is right: he looks great to my eye, but his times are slowing and, Larsen assures me, he's not coming off the ground with quite the same *pop.* "Fatigue is setting in. He's not ready to race yet," he says. "He's looking good, but we've still got three more weeks."

When Meb crossed that finish line, he became the oldest man to win the Boston Marathon in more than 80 years. To appreciate the significance of that, it helps to know something about the biology and physiology of aging. There's a reason so many spectators, expert and otherwise, assumed one of Meb's younger rivals would overtake him, even though we all read "The Tortoise and the Hare" as kids.

Getting older brings changes to just about every system of the body at every observable level, from the ones you can see with the naked eye to the ones we can only detect with an electron microscope. The vast majority of those changes are unwelcome. There are, of course, the purely cosmetic ones: your hair turns gray as the melanocytes in your follicles stop producing pigment; your skin gets wrinkly as the cells that produce elastic collagen wind down and the fat deposits underneath it wither. (If you're lucky enough to go attractively silver without getting wrinkles, you have a bright future as an actor in pharmaceutical commercials.) Then there are the changes that reflect a deeper process with implications for athletic performance. The average American gains about one pound a year from early adulthood through middle age. According to the Centers for Disease Control and Prevention (CDC), the fastest gains occur in one's 20s; epidemiologists believe a key factor is people who played sports in high school and college giving them up to focus on work. (This may not be as true for younger generations of adults as it was for their parents, but it's still true.) Sometime after age 50, the trend reverses itself and the pounds start coming off. But the explanation for that turns out to be more bad news. It's called sarcopenia, which means muscle wasting, and for most people it begins gradually around age 40. After 50, the rate of muscle loss exceeds the rate of fat gain. By 70, you will likely have lost more than 15 percent of your lean muscle mass, and from that point on, unless you're doing something special to keep it, you'll probably lose 30 percent of it each decade for as long as you're around.

As you can imagine, losing muscle means losing strength. But it's not the only reason you get weaker. A researcher named Stéphane Baudry of the Université Libre de Bruxelles discovered in 2014 that older people performing a muscle contraction experience "more coactivation of the antagonist muscle," resulting in less net torque on the joint. "Coactivation of the antagonist muscle" means if you were attempting to do, say, a biceps curl, your triceps would be firing at the same time, resisting the action. In other words, as though getting older didn't cause enough problems, your muscles literally start working against you.

Your most important muscle also loses some of its vigor: even in the healthiest humans, the maximum attainable heart rate falls steadily throughout one's life span, starting, for most of us, around 200 beats per minute in early adulthood and falling by about 1 BPM per year thereafter. Lung capacity diminishes, too, as the air sacs that make up the lungs become less elastic in response to a lifetime of pollution, and the rib cage, like the rest of you, loses some of its suppleness. Add all that up — a heart that doesn't pump blood as fast, lungs that don't deliver as much oxygen to the bloodstream, and smaller muscles that don't pull as much out of it — and you get a steady downward march in what sports scientists call VO_2 max, the body's maximal ability to use oxygen during activity.

Continuing the tour: your bones lose density, making them more susceptible to fracture. This is especially a problem for females past the age of menopause; around one-third of women over age 50 have osteoporosis in the hip, spine, or forearm. Once broken, your bones take longer to knit the older you get. In fact, any sort of injury, from muscle strain to ligament sprain to a simple cut or bruise, heals more slowly. Biologists don't understand all the reasons for this but believe it has something to do with decreased production of growth hormones and deteriorating function of the specialized stem cells that carry out the repairs. Slower healing is also the rule for the trauma that occurs in muscles in response to hard training. While microscopic tears are what trigger the muscle adaptation you're seeking when you lift weights or run, it takes longer for those beneficial changes to manifest, meaning older athletes need more time to recover between hard workouts. It's also harder for your muscles to convert the protein you eat into new fibers thanks to a phenomenon called anabolic resistance.

After 45, osteoarthritis — or painful inflammation of the bones at the joints — becomes much more common. This happens as the cartilage that acts as a shock absorber in those joints, particularly in the knees, wears down and the cells that help it regrow get worse at their job — again for reasons that aren't totally understood. The shocks that cushion the vertebrae of your spine take a beating, too. By age 50, more people than not have at least one bulging intervertebral disk, even if they don't experience any symptoms. The relationship between disk abnormalities and pain isn't straightforward — it's common to have either one without the other — but most people will suffer low-back pain at some point in their adult lives. As you exit your 40s, your risk of a herniated disk shrinks. That sounds promising. Unfortunately, it's because the disks themselves are also shrinking, which not only predisposes you to new types of pain but explains why you can expect to get a little shorter with each passing decade.

Your nervous system is changing, too. Human reaction times are at their best around age 24 and drift upward from that point on. That has to do with the slower speed at which nerve signals travel: as the protective casings of protein around peripheral nerves degrade, they don't conduct impulses as efficiently. The additional "noise on the lines" is one reason the simple act of balancing requires more conscious effort in the elderly. (Or the not-so-elderly: try standing on one foot with your eyes closed and see how long you can last before tottering over.) Generally speaking, the senses grow duller. There aren't a lot of sports that call for a keen sense of taste or smell or require the ability to hear high-pitched frequencies. But the ability to read the rotation of an oncoming baseball pitch or tennis serve is a plenty useful thing (and that much more useful if your reaction times aren't what they once were). At least the vision change most associated with midlife — the need for reading glasses to see up close, typically starting around age 40 — is more of an issue for coaches than jocks.

When you zoom in to examine the mechanisms that drive all these changes, all sorts of intricate interconnections and feedback loops emerge. Many symptoms of aging are linked to decreased hormone levels, particularly the hormone testosterone. The less testosterone you have, the harder it becomes to retain and build skeletal muscle (i.e., all the muscle that's not part of your circulatory system or digestive tract). Skeletal muscle burns a lot of calories; as you lose it, your metabolism slows, meaning any calories you consume are more likely to end up as fat. And fat secretes the hormone estrogen and proteins that promote chronic inflammation and insulin resistance. As the writer Bill Gifford put it in *Spring Chicken,* his 2015 tour of antiaging science, "Aging makes us fat, and then our fat makes us age."

That's a lot of bad news to digest, I know. (Speaking of digestion, that gets worse, too. Good luck chugging that bedtime protein shake without triggering a bout of acid reflux.) Here's the good news: most of the major changes enumerated above can be attenuated, delayed, or reversed through frequent and vigorous exercise. It won't keep your hair dark or prevent you from needing bifocals, of course. But the most pernicious symptoms of aging — cognitive decline, muscle wasting, bone thinning, cardiovascular impairment — just don't happen in the same way in people who work out hard and often.

You can probably tell a masters athlete from his collegiate counterpart by adding up the wrinkles. But if you were able to look inside and see what's happening on a molecular level — observe which of their genes are switched on or off, measure what hormones and other proteins are circulating in their blood, quantify the coordination of signals across their nervous systems — you would likely find the two have more in common than that masters athlete does with a sedentary person his own age.

One landmark study published in 2007 looked at what was happening inside the muscle cells of healthy adults before and after they started resistance training. The scientists specifically focused on the mitochondria, the components of each cell that function as power generators. Mitochondria have their own genomes, distinct from the 26 chromosomes in the cell's nucleus. Biologists don't understand all the mechanisms involved, but they do know that something about the way these mitochondrial genes get expressed — that is, decoded from DNA and translated into the production of new proteins — plays a central role in cell death and aging. The researchers put 14 older adults on a six-month exercise program and biopsied their muscles before and after so they could see what proteins were being manufactured and how those protein profiles compared with the muscle cells of a control group. As they put it in their paper, "Following exercise training the transcriptional signature of aging was markedly reversed back to that of younger levels for most genes that were affected by both age and exercise." Other scientists have linked the proteins produced after exercise to a lower risk for diseases of aging like Alzheimer's and atherosclerosis. In other words, at a basic biochemical level, exercise restores the hallmarks of youth.

This is equally true when you pull the camera way back and look at things through the lens of population health. A few years ago, researchers at the University of Science and Technology in Trondheim, Norway, developed a concept they labeled "fitness age" to quantify how a person's cardiovascular fitness translates to his or her risk of death from any cause, from heart attacks to falling anvils. They were working off data from a population study of 55,000 adults, which suggested that a person's VO_2 max is a remarkably good predictor of longevity, more accurate than blood pressure, cholesterol levels, or even smoking history. The Norwegian findings indicate the typical elite endurance athlete in her 50s likely has a physiological age somewhere in her 30s; that is, she will die 20 to 30 years later than conventional life expectancy tables would predict. That agrees with a 2011 study, which concluded that athletes in their 70s who maintain a high-quality training regimen have physiological measurements similar to athletes in their 40s.

Heart disease is the number 1 cause of death worldwide, so it doesn't require a PhD to understand how strengthening your cardiovascular system can improve your chances of long life. What's a little more mysterious is why exercise seems to have such a powerful effect on brain health. Multiple studies have shown even moderate forms of physical activity, like walking, help preserve the volume of gray matter in older brains and substantially reduce risk of Alzheimer's disease. Intense exercise is even better; regularly pushing yourself to physical exhaustion is correlated with being a so-called superager.

The more we've learned about how our bodies respond to exercise, the more we've come to understand how even aspects of it we once thought of as potentially harmful are beneficial. What we once thought of as trade-offs are actually win-wins. For a long time, it was considered self-evident that running, however salubrious it may be for the heart and lungs, is hell on the cartilage in your knees. For people over 40, low-impact alternatives like swimming or cycling were considered preferable. In the last few years, however, researchers have found increasing evidence pointing in the opposite direction. It turns out that applying force to cartilage in a repetitive pattern is what stimulates cells called chondrocytes to manufacture more cartilage. Running is one activity that produces this pattern of force, called cyclic loading; jumping rope, skiing moguls, and squatting with weights are other ways to achieve it. Swimming, cycling, and walking don't have the same effect. Cyclic loading is thought to explain why regular runners have a lower incidence of knee osteoarthritis than nonrunners, even when you control for the fact that runners tend to be thinner.

Donald Trump famously tells people he doesn't exercise because he believes the body is like a battery — the more you use it, the sooner it runs out. But the correct metaphor is something you would think Trump would understand. In reality, the body is a bank, and exercise is a loan that gets repaid with interest.

As it happens, a great deal of what we know about the physiology of exercise and how athletes age comes from the study of endurance athletes like Meb. Partly, that's a matter of logistics. Compared with athletes from other sports, distance runners, cyclists, and rowers make relatively easy study subjects. It's hard to reproduce what a soccer forward or a mixed martial artist does in a laboratory setting; there are just too many variables involved. Endurance sports, on the other hand, make for natural experiments. For all the strategy and psychology involved in winning a race, the math is pretty simple: if I know where you started and finished and how much you weigh, I can figure out how much work you performed during the race, more or less. And if I can stick you on a treadmill or a stationary bike and put a mask over your face that captures your exhalations, I can measure how much oxygen you're consuming and, while I'm at it, maybe even draw a bit of your blood and study your lactate and hormone levels.

But we also know so much about endurance athletes because so many of the pioneers of sports science were themselves distance types who created an entire field of study largely out of a desire to understand the thing they loved. The grandfather of endurance studies is David Costill, the longtime director of the Human Performance Laboratory at Indiana's Ball State University. As an undergrad at Ohio State in the 1950s, Costill was a sprinter on the swim team. After graduation, he continued to work with swimmers as a coach while teaching high school science, but he was unsatisfied with his career path and went back to school to earn a doctorate in physiology and physical education. By the early 1960s, Costill had switched over to coaching cross-country and taken up running marathons himself while finishing up his doctoral research at the State University of New York at Cortland.

By that point, he had spent enough time around elite endurance athletes to realize he wasn't like them. Some mysterious mix of innate characteristics separated the champions from everyone else. Costill, whose best marathon time was a pedestrian 3:15, burned to understand that mix. "I wanted to know why all these guys were faster than me," as he put it. He also wanted to know what training methods would allow a plodder like himself to make the most of his limited potential, but he lacked the resources to pursue answers to any of those questions.

Then, one day in 1961, Costill's life changed in a way that would seem like a cheap plot device if it happened in a movie. He got an envelope in the mail with no return address; inside was a newspaper clipping advertising a job opening for someone to head up Ball State's brand-new Human Performance Lab. He was never able to learn for sure who sent it. Whoever it was, he or she couldn't have nominated a better candidate. In the course of his 50 years at Ball State, Costill has published over 400 papers on the physiology of exercise, many of them considered to be among the foundational studies in the field. He has investigated everything from how long before race day runners should taper their training to what happens to astronauts' bones and muscles during months in zero gravity. No one has done more to elucidate what's going on inside people's bodies when they jog, swim, or bike. (For better or worse, he has also helped pioneer the practice of industry-funded sports science. One of his first studies at Ball State was a look at dehydration in runners, paid for by a then-new beverage brand called Gatorade.)

Throughout it all, Costill continued to race, competing in marathons for 20 years and then switching back to swimming after his 70-mile-a-week regimen proved too much for his knees. (Apparently his chondrocytes didn't get the memo about the benefits of cyclic loading.) As a masters swimmer in his 40s and 50s, Costill found himself posting faster times than he had as an undergrad. His scientist's mind became increasingly preoccupied with questions about how age influences performance. Early in his career, it had seemed like a settled matter: physical ability peaks in the late teens or early 20s and declines rapidly from there. But maybe it only seemed settled because so few people were even attempting to challenge it. "When I started in this field, you were a real oddball if you were out running and you were over thirty years old," he told me. Since then, Costill has seen more than enough data to convince him that endurance athletes peak later and age more slowly than anyone once suspected. "That whole thing has sort of moved forward," he says. "I see people doing lifetime bests well into their forties."

The ideal way to understand aging is with a longitudinal study, which follows a set of subjects over a long span of time. Having been at the same lab since the 1960s, Costill is better positioned than most to conduct such a study. In the mid-1990s he and his protégé, Scott Trappe, assembled a group of 54 runners of various achievement levels, from world-class to recreational, for physical testing. Costill had tested all of them 25 years earlier at the Human Performance Lab, allowing for true apples-to-apples comparison. "As you might expect, several changes in lifestyle and body type had occurred over the years," Costill and Trappe drily noted in their 2002 book *Running: The Athlete Within*. (Trappe eventually succeeded his mentor as director of the lab and a leading light in sports science research; Costill serves as professor emeritus.) For those who had kept up with running in the intervening years, however, the onset of middle-aged spread was in fact minimal: competitive runners in the group had gained an average of only 4 pounds in 25 years, while those who ran for fitness had put on 10 pounds. At the other end of the spectrum, former marathoners who had given up running had gained an average of 30 pounds, or about the norm for a typical American.

More to the point, the serious runners in their group had maintained their fitness characteristics to a remarkable degree. When weight changes were factored in, masters-level runners in the follow-up study had experienced only a 7 percent reduction in VO_2 max. Several tested within 2 percent of their values from 25 years earlier. This is consistent with race-time data. In track and field, masters events are open to anyone over 40. By that age, distance runners can expect their times to increase 4 to 8 percent. That translates into a mere six-minute difference between the world-record marathon time of 2:02:47 and the masters record of

2:08:46. The erosion of VO$_2$ max and race performance remains fairly moderate, if steady, until about age 70, when it accelerates dramatically.

If you're a marathoner like Meb, hoping to bring home an Olympic medal, a six-minute gap is nontrivial, of course. Then again, recall that he went into Boston spotting his rivals a five-minute advantage on paper. Certainly runners in other events would kill to be that competitive at 40. Sprinters pretty much never are. In the 100 meters, the difference between world record and masters record is more than 7 percent for men and 13 percent for women. The reason has to do with muscle composition. As research by Costill and Trappe (among others) has shown, muscle cells come in different types, often referred to informally as slow twitch and fast twitch. As their name suggests, slow-twitch (or type I) muscle fibers contract relatively slowly but are also slow to fatigue. Fast-twitch (type IIa or IIx) fibers both contract and exhaust themselves more rapidly. Very roughly speaking, the slow-twitch/fast-twitch divide is analogous to the one between aerobic and anaerobic respiration. Everyone's muscles are a mix of the two, but there's a lot of variability in the proportion we're born with, and there's not much anyone can do to alter their individual mix. At the elite level, athletes whose sports require bursts of explosive movement tend to have more fast-twitch fibers and those whose sports demand endurance have a higher ratio of slow-twitch. Olympic sprinters like Usain Bolt can have leg muscles comprising 80 percent fast-twitch fibers.

Here's the key: performance in fast-twitch-dominant athletes deteriorates more rapidly with age. (Raw physical performance, that is. Other aspects of performance often improve, but we'll get to that.) Sprinters typically achieve their career bests five years earlier than marathoners, around 25 versus 30. That's because sarcopenia claims fast-twitch fibers at a higher rate, leaving behind more slow-twitch ones. It's not entirely clear why this is. One hypothesis is that, since fast-twitch fibers have a higher threshold for activation, they don't get called into action as often and thus are more prone to atrophy through disuse. Those of us who aren't elite athletes most likely amplify this effect by modifying our activities as we age, consciously or unconsciously, in response to injuries or lifestyle factors. If you've swapped out your weekly basketball game for a session on the elliptical, you've nudged your muscle composition toward more slow-twitch. "Use it or lose it" is a useful motto in all areas of fitness, but when it comes to muscle fibers it's quite literally the case.

None of that explains what Meb is doing here in Mammoth, a seven-hour drive from his home and family in San Diego, three weeks before a major race. He's here for the thing I noticed on the drive in: the thin mountain air. The benefits of training at high altitude weren't generally understood in the endurance-sports world before 1968. That was the year the Summer Olympics were held in Mexico City. At 7,300 feet of elevation, it's about as high as Mammoth, and it showed in the disappointing finish times of the runners from lower-lying regions.

More than 30 years later, however, the average American cross-country coach remained reluctant to incorporate altitude training, and people like Bob Larsen were getting impatient. Larsen knew that 95 percent of medals won in international competitions went to runners who trained at altitude. He believed American runners' reluctance to do so went a long way toward explaining why the country hadn't produced a dominant marathoner in decades. "A lot of people respected in the sport would say you don't need to train at altitude unless you're going to race at altitude," he says. "You have to put yourself in their minds: if altitude really worked, it was a disadvantage for most coaches because they're based in colleges that are at sea level, so there's a built-in bias that altitude is not the solution."

At a seminar he attended in 1999, Larsen had seen research suggesting people who live near sea level can obtain the strongest benefits from spending a few weeks living and training at between 7,000 and 8,000 feet. At that elevation, the low oxygen density triggers the kidneys to produce the hormone erythropoietin, or EPO, which in turn induces the bone marrow to manufacture more red blood cells. (You've probably heard of EPO. It was one of the drugs Lance Armstrong used during his run of seven consecutive Tour de France titles, all since rescinded.) The hemoglobin in red blood cells is what carries oxygen from the lungs to the muscles; the more of it you have, the harder your muscles can work without going into anaerobic oxygen debt. In fact, Larsen knew that even the Ethiopian marathoners who lived at 7,000 feet were making use of altitude training, trekking up to 10,000 feet for extended periods to goose their own red-blood-cell production.

He also knew Mammoth, having visited it regularly on ski trips since the 1960s. Larsen had been bringing his UCLA teams up for training weeks for years when, in 2001, he and another coach, Joe Vigil, founded the Mammoth Track Club. The idea was to convince a few elite runners to embrace altitude training and then use their results to win over the rest of the American running establishment. And that's pretty much how it went down. Among the group's charter members were Meb and marathoner Deena Kastor. In 2004 both spent months leading group runs around Mammoth in preparation for the Athens games. Then they went to Greece, where Meb won silver and Kastor won bronze. It had been 28 years since an American man had medaled in marathon at the Olympics, and 20 years since a woman had. "When we came back from Athens, it was a lot easier to convince people Americans should train at altitude," Larsen says. And it's gotten even easier since. After American distance runners took home seven medals from the Rio Olympics, the country's best showing in over 100 years, the president of USA Track & Field told *Run-*

ner's World it was largely a product of the team's increasingly sophisticated use of altitude.

Elite athletes don't need extra incentive to adopt anything that
can help them win, but in the case of altitude training, there may
be an ancillary benefit even Larsen wasn't anticipating when he
started championing it, one that helps explain why Meb remained
a top marathoner at 41 (and why Kastor, who is two years older,
continues to win races, although a strained glute forced her to
drop out of Olympic trials). Training at elevation isn't just effica-
cious; it's also efficient. "When you train at high altitude, it takes
less mileage and it takes less intensity," Larsen says. For Meb, that
means tempo runs at a pace of five minutes per mile, knowing that
translates to a 4:30 in New York or Boston or San Diego. "You're
not beating up on your body quite as much as you would to get the
same effect at sea level."

Larsen's observation is based on experience, not data, but it
makes a lot of sense. Meb has run thousands and thousands of
miles in Mammoth over the course of his career. Shaving 10 percent of the pounding off each of those miles has to add up over the
long haul, doesn't it? "Being able to put in a lot of miles just a little
slower, not having to push quite as hard, I think does have long-
term advantage," he says.

By now, we're parked on the shoulder of the road, not far from
a hot springs, and close to a stream where Larsen says bald eagles
often fish. After tacking on a three-mile cool-down run on another
dirt track, Meb is sitting on the ground stretching his hamstrings
and ankles with a strap he retrieved from the truck. I ask Meb:
What does it feel like to run here, for someone who's never done it
before? Would I suck wind?

You should try it, they both tell me. "You'd have to start out really slow," Larsen cautions. "If you started out at your normal rate
at sea level, you'd go into oxygen debt. At altitude you can't back off
enough to get air in." We're at about 7,100 feet now, he says; to get

the full effect, he suggests I drive up to the area of the Mammoth Lakes themselves, at 8,500 feet.

Bidding them farewell, I do that. I'm a twice-a-week kind of runner and it's been four days since my last outing, so my legs ought to be plenty fresh. Heeding Larsen's advice, I start down the path that loops around the Twin Lakes at what I reckon is a little slower than my usual pace. I feel the altitude immediately. It's not so much a shortness of breath as a feeling of all-over heaviness I have to fight through, almost as if the air is thicker, not thinner. I dial back the pace still more and focus on the scenery, which, conveniently, is unbelievably beautiful. The path winds through stands of aspen and fir; to my left, a hawk screams as it plunges down the face of a cliff toward the sparkling surface of the lake. At the turnaround point, my lungs still feel not-bad, but even at this slower pace, my body is just not working the way I'm used to. Instead of my feet *pop-pop-popping* off the pavement like Meb's, mine seem to be spending extra time hanging out there, like two drunks taking turns leaning against a wall.

I'd brought my running clothes to Mammoth with the vague notion that I might have been able to persuade Meb to let me tag along with him for his warm-up or cool-down. I now realize how far from reality that idea was. On a great day, I'm a seven-minute-mile guy back home. Up here I'd be lucky to hold that pace for the length of a Ramones song. After 25 minutes I call it a day with a whole new respect for Meb. And for my Subaru.

2

FRESH IS THE NEW FIT

*A Better Way to Think
About Conditioning*

Jaromir Jagr doesn't do anything halfway. Four months into his 25th NHL season, the 44-year-old Czech hockey star, a forward for the Florida Panthers, decided it was time to quit drinking coffee. A devout Christian, Jagr picks something different every year to give up for Lent. The previous season, it was Diet Coke. Although he had been pounding them at the rate of five or more a day, Jagr kicked the habit easily enough, to the point that, once Easter had come and gone, he decided he no longer liked the taste and declined to resume drinking it. Coffee was different, though. "He just loooooooves coffee," says Tommy Powers, the Panthers' strength and conditioning coach and, whether he likes it or not, Jagr's near-constant companion. "He would drink it all day long. He must have been having at least ten to twenty cups a day." Powers remembers returning to Fort Lauderdale after a road trip in the middle of the night and pulling into a gas station on the way back from the airport. Although it was after 2:00 a.m., Jagr ran inside for a pick-me-up to tide him over until bedtime. "He was basically addicted," he says.

But Jagr kept his vow. What's more, he did it cold turkey. "I felt awful the first few games. I felt like I had no energy at all," he told a reporter three weeks into his java fast. No wonder: caffeine is the world's most widely used performance-enhancing substance. Its effects are so potent — one meta-analysis of its use by endurance racers found a 3 percent improvement in finish times at optimum dose levels — that the World Anti-Doping Agency for years set a limit on how much athletes could have in their systems. (WADA dropped the limit in 2004, essentially acknowledging that we're all dopers when it comes to coffee.) Caffeine withdrawal is so debilitating, the American Psychiatric Association added it to its official list of recognized conditions in 2013, noting that heavy users can experience effects for up to three weeks. Jagr had been mainlining at least 1,000 milligrams a day of stimulant; that he felt like garbage without it was to be expected. What wasn't predictable was how he responded on the ice: in the 20 games he played during Lent, he put up 18 points, well above his average rate of scoring for the season.

To say that Jagr — "Yags" to his teammates — is a man of unusual discipline and unique temperament is a considerable understatement. Here are a few of the things he likes to do on a daily basis: Skate sprints up and down the ice after games, wearing a 45-pound weighted vest and additional weights on his ankles. Practice his shot with a six-pound medicine ball, slapping it into the wall hundreds of times in a row. Load up a sled with iron plates and run sprints towing it. Attach the end of his stick to the cable on a weight machine to simulate the feeling of a defender trying to muscle it off the puck. Pedal as hard as he can on an exercise bike, going one minute on, one minute off, until he can't keep going anymore. None of this is mandatory. In fact, Powers might well prefer if Jagr scaled it back a bit, since Jagr often calls him in the middle of the night and asks him to supervise these extracurricular sessions. Even during games, Jagr will sometimes try to squeeze in extra conditioning work, impressing Powers into service for medicine ball throws or resisted sprints between periods.

When Jagr is finally spent, Powers has him stretch out his six-foot-three, 220-pound frame using a technique called Ki-Hara Resistance Stretching. With Jagr seated in butterfly position, the soles of his feet pressed together, Powers pushes his knees down while Jagr uses his adductor muscles to try to draw them together. This achieves what Powers calls "eccentric strengthening," eccentric referring here not to Jagr's proclivities but to the act of lengthening a muscle while it's contracting, as when you're slowly lowering yourself from a chin-up. (When a muscle contracts while shortening, that's called concentric.) The swimmer Dara Torres, who won three Olympic medals at age 41 and swam in five different Olympiads, is among the well-known proponents of Ki-Hara. The idea behind it is that flexibility without strength only invites injury. You should be able to generate force from anywhere within your range of motion. "This way, if they get into an elongated position on the ice that's kind of unprecedented, that they don't expect to happen, the muscle is still strong in that position. You're able to contract out of it," says Powers. Even when Jagr is stretching, he's getting stronger. (Ichiro Suzuki, the 44-year-old Major League Baseball hitter, follows a training program based on similar principles, except instead of having a trainer resist his eccentric contractions, he uses specialized Japanese-made machines. A circuit of seven machines is the only strength work he does, and he does it as often as three times a day. Suzuki has almost never been injured.)

It all sounds extreme, but it's hard to argue with the results. Just playing in the NHL at 44 is a remarkable accomplishment. Excluding goalies, only two other players have ever done it: all-time great Gordie Howe, who retired shortly after his 52nd birthday in

1980, and Chris Chelios, who made it to 48. Physiologically, hockey could hardly be more different from running marathons. Players cycle on and off the ice in "shifts" of 30 to 90 seconds, skating furiously the whole time they're in the game. By the end of a shift, their hearts are beating at more than 90 percent of their maximum capacity, lactic acid is fast accumulating in their muscles, and they're in severe oxygen debt. Hockey requires the exact types of strength and fitness — explosive, fast-twitch, anaerobic — that decline fastest with age. And that's before you get into all the other stuff that happens in a hockey game that takes a toll on the body: the brutal checks into the boards, the fistfights, the almost-inevitable hip and groin injuries. According to one study, the average NHL team loses 242 man-games to injury per year, the highest injury rate of any major sport. No wonder a 2015 literature review in the journal *Sports Medicine* concluded the typical hockey player peaks in performance by age 28.

But Jagr hasn't just survived, he's remained a star, leading the Panthers in scoring, taking them to the playoffs, and becoming the oldest player ever to score 60 points in a season. Along the way, he's put his imprint on the culture of one of the league's younger teams, convincing up-and-comers like 20-year-old Finnish center Aleksander Barkov to emulate his unconventional fitness regimen and even some of his more idiosyncratic practices, such as using different types of music to open his chakras depending on his energy level. "Those younger guys will listen to just about everything he says with open ears and bright eyes," says Powers.

One thing they won't do is outwork him. The NHL's collective bargaining agreement mandates that players get at least four days off per month during the season. Jagr disdains days off. He sees rest as an enemy, one to which no quarter must be given, lest sloth and idleness seize the opening to undo all his hard work. Maintaining momentum is everything. "It's like a truck, a heavy truck," he said at the end of the season, shortly after signing a new one-

year, $4 million contract. "When you're going, you kind of go —
and fast. But once you stop, it's tough to start again." It sounds exhausting, like being Keanu Reeves in *Speed*, but on ice skates and
every day for 25 years. Jagr likes it. "Physically, I don't get tired,"
he told the *South Florida Sun-Sentinel*. And that was *after* quitting coffee.

Jaromir Jagr is sui generis, but he's also a familiar type in the
sports world: the aging athlete who believes the key to longevity
is training harder with each passing year to compensate for what
time takes away. Every professional team seems to have a veteran
who boasts that he's the fittest guy on the team and is happy to
prove it to any youngsters gunning for his roster spot. Training for
what would prove to be a championship season in the fall of 2015,
LeBron James humblebragged to reporters that he had gone "a
little bit crazy with my workout regimen," stepping up his schedule to three workouts a day as his 31st birthday approached. "Did
I get enough rest? I don't think so . . . I could definitely use a couple more months off," he said. "But I definitely improved." A 2016
ESPN profile of Minnesota Vikings running back Adrian Peterson
detailed the 31-year-old's heroic summer program, which includes
running stacked sets of 300-meter sprints in the East Texas heat
before heading inside for hours of weight training. Featured ballcarriers like Peterson almost never play well into their 30s. But,
noted the writer, "[in Peterson's] mind, he'll be successful longer
because he works harder."

   U.S. soccer midfielder Carli Lloyd credits her late-career as-
cendance — she was 33 when FIFA named her the world's best
woman soccer player — to a work ethic that includes doing 400
push-ups and 800 sit-ups every single day and never taking a day
off, not even Thanksgiving or Christmas. James Galanis, Lloyd's
longtime coach and personal mentor, calls this the "brushing her
teeth" method, meaning she'd as soon skip her daily hour of calis-

thenics as walk around all day with morning breath. During the 2012 Olympic Games, he recalls, Lloyd decided she wasn't getting enough fitness work just working out with the team, so at six thirty in the morning she set out water bottles on the lawn of the house she was renting and used them as cones for sprint drills.

Andre Agassi, who became the oldest man ever to be ranked number 1 by the World Tennis Association in 2003, when he was 33, has a similar narrative. In his autobiography, *Open*, he described the aftermath of the punishing workouts his strength coach, Gil Reyes, designed to keep his aging, injured body in shape for five-hour Grand Slam matches: "Easing into my car at dusk, I often don't know that I'll be able to drive home. Sometimes I don't try. If I don't have the strength to turn the key in the ignition, I go back inside and curl up on one of Gil's benches and fall asleep." Agassi believes his superior fitness, particularly the conditioning of his legs and core muscles, helped him compensate for a congenital back problem, outlast younger opponents, and win four of his eight Slam titles after age 29, the age by which most tennis players have begun to fade.

It's not just a matter of avoiding performance decrements. In the summer of 2011 I traveled to Wisconsin for the opening of the Green Bay Packers' training camp, six months after their victory in Super Bowl XLV. As the first practice of the season ended, the players left the field in twos and threes until the only one remaining under the lights was 36-year-old wide receiver Donald Driver. With moths drifting around him, Driver stood in front of a JUGS football-throwing machine, his jersey hiked up to show off his hyperarticulated abs, catching one perfect spiral after another.

Despite playing most of his career in the era before the advent of so-called defenseless-player rules banning the most dangerous types of tackles, Driver had missed only six games in his previous 12 years as a starter. That made him an extreme outlier in the NFL, where more than 90 percent of players will sustain a major injury over the course of their careers and the average wide receiver gets injured once every 250 snaps. Driver played close to 10,000 snaps in his career, enough to make him the team's all-time leader in receptions.

It's an even more impressive record when you consider the particular demands of Driver's job. The biggest, fastest wide receivers usually operate along the sidelines, where they rarely get tackled in head-to-head collisions. More quick than blazing fast, lacking the big frame of a number 1 receiver, Driver made his living catching short passes in the shallow middle of the field, where hulking linebackers and headhunting safeties can launch themselves into ballcarriers at a dead sprint. Some star wideouts intentionally dog it on these passing routes, making a "business decision" to avoid catches that might come with season-ending consequences. Driver says he never considered that an option. During his rookie season in 1999, he got a piece of advice from Antonio Freeman, then the Packers number 1 receiver, that stayed with him. "Free always told me, 'If you want to make it in the National Football League, you have to go across the middle,'" Driver says.

Ask Driver to explain his statistically improbable ability to stay healthy and he credits two individuals: God and himself. "I trained my body to withstand anything," he says. While most veterans start to ease up after they pass 30, "I just continued to work harder and harder." Such was his dedication, before his final season in the NFL, he cut his body fat down from 4 percent to 2 percent, a level usually attained only by marathoners and cyclists. "I always tell the young guys, 'When you're sleeping, I'm working,'" he says. Driver's inspiration in all this was Jerry Rice, who played 20 seasons and is generally considered the best receiver ever to have played the position. Like Driver, Rice didn't have exceptional straight-line speed, but until the end of his career, he was known for a maniacal devotion to fitness, including a six-day-a-week off-season program of daily wind sprints, longer runs, and weights.

Some of Driver's methods, like Jagr's, can seem questionable at first blush but on closer inspection betray a savant's knack for sports science. Throughout his career, Driver refused to drink water during games or practices, believing it slowed him down and made him soft. "After I get done with practice I drink as much as possible but while I'm practicing I train my body for endurance," he said. That sounds like superstition; certainly it runs counter to the modern, Gatorade-sponsored mania for hydration. But in fact, studies show training in a state of mild dehydration, when done correctly, does promote endurance by causing the body to increase blood plasma volume.

Then there's Driver's workout. Although he reluctantly embraced the squat rack after he turned 30, Driver was never a fan of lifting heavy. Working out on his own during the off-season, he stuck to a regimen centered on sprinting, agility, and footwork drills such as running through ladders and between cones. Driver performed these exercises in sessions of 40 minutes, with minimal pauses in between. Recent studies have shown high-intensity, low-dose interval training of this sort has similar effects on VO_2 max and other fitness markers as much higher volumes of lower-intensity exercise. There's also a body of evidence that plyometric training, which involves jumps and other explosive movements like the ones Driver practiced, can reduce the risk of injury by improving balance and proprioception, or the awareness of one's body in space. The data is strongest when it comes to ankle sprains and other lower-leg injuries of the type most likely to sideline a wide receiver. Plyometrics have been used in American athletic training since the 1970s, when a track coach named Fred Wilt coined the term, but they've usually been seen as a side dish. For Driver, they were the main course.

Driver wasn't thinking specifically about avoiding injuries when he made up his off-season workout plan. "I just did what I had to do to keep my body in great shape and to be a specimen," he says.

When he retired in 2013, "I left the game feeling healthy, no major injuries," he says. Again, that makes him unusual: more than 80 percent of former players say they experience pain throughout the day as a legacy of playing football. Calvin Johnson, the dominant wide receiver whose Detroit Lions played Driver's Packers twice a season for years, retired at 30, saying it had become difficult for him to walk when he got out of bed in the morning. In a 2009 survey of retired NFL players conducted by the University of Michigan's Institute for Social Research, 84 percent of players of Driver's generation said injuries were a major factor in forcing them into retirement.

Even in sports that don't subject players to repeated ballistic collisions, injuries are arguably the driving force in shortening careers. Every sport has its characteristic injuries of trauma or overuse: ankle and knee arthritis in basketball, elbow tendinitis in tennis and golf, rotator cuff impingement and disk degeneration in swimming. Michael Johnson, who won four Olympic gold medals as a sprinter and now owns a performance center outside Dallas, told me injury, not age-related slowdown, is almost always the biggest factor in determining when athletes retire: as breakdowns get more frequent and take longer to heal, it simply becomes too difficult to do the amount of conditioning necessary to win.

In chronically hurt athletes, recurring injuries can in fact represent a form of accelerated aging. As noted earlier, there are specialized stem cells that carry out the repairs in damaged muscles, and scientists believe a decline in the functioning of those cells is one reason our muscles (and bones and tendons and ligaments) mend more slowly when we get older. They also believe there may be an upper limit to how many times those cells can perform their jobs, which require them to reproduce by division. Injure the same tissue too many times and it just won't heal right anymore. That part of your body has become old, even if the rest of you isn't.

It's clear there's a relationship between fitness and durability.

While theories on why athletes get injured abound, just about everyone agrees playing a sport without being fit enough to meet its basic physical demands is an efficient way to get hurt. But it doesn't necessarily follow that the fittest individuals, the Jaromir Jagrs and Carli Lloyds and Donald Drivers, are the least likely to injure themselves. It's a chicken-and-egg problem: Do they stay healthy because they're so fit, or are they so fit because they stay healthy enough to train so hard? What if we have the whole thing backwards?

For a guy whose life's work is helping people avoid needless pain and debility, Raymond Verheijen is rather unpopular with his peers. That's very much his own doing. The Dutch pride themselves on being a people who value honesty above politeness, and Verheijen possesses a full measure of his national gift and then some. On a typical Saturday between September and May, he can be found on Twitter cataloging the stupidity and pigheadedness of Europe's most celebrated soccer coaches. One particular target of his scorn is Arsène Wenger, the Frenchman whose reign as manager of London's Arsenal Football Club has spanned more than 20 years. Thanks to Wenger, Arsenal has long been distinguished by its stylish, attacking brand of play in a league where punishing defense and opportunistic offense are the norm. Another Arsenal calling card has been its almost unbroken history of lousy injury luck, with season after promising season deflated by some key player's long absence. Except Verheijen thinks luck has nothing to do with it, and he seldom misses an opportunity to say so. "The moment Arsene Wenger accepts the structural Arsenal injury chaos must have something to do with him it will be the start of the solution," he tweeted on the opening weekend of the Premier League's 2016–17 campaign, after Aaron Ramsey, a young Welsh midfielder, limped off with yet another hamstring pull. "As long as Wenger keeps blaming external factors, nothing will change."

Verheijen used to share opinions like these with reporters but came to believe they were in the pockets of teams, too concerned with preserving their access to acknowledge the obvious but unpopular truth. Now he reserves his views for social media and for the seminars he gives through his World Football Academy, which offers training for soccer coaches. "What is important when you interview me is you understand I am not subjective but objective," he tells me. "I don't give my opinion. I just describe what is happening."

What is happening, in Verheijen's objective view, is that most of the world's soccer teams are training their players the wrong way, damaging their bodies, undermining their performance, shortening their careers. They're doing so out of a misguided belief that the way to prepare players for the rigors of a physically grueling sport is to make their practice sessions and fitness work proportionately more grueling so the games will seem easy by comparison. By knowingly subjecting players to training conditions that exceed any game in intensity or duration, these teams are creating perpetually exhausted athletes who stand a better chance of exiting any particular game on a stretcher than they do playing their best in it. And the problem is particularly acute when it comes to older players, for whom the cycle of overtraining and injury only serves to hasten the end.

Like most people you meet in the sports world, Verheijen started out with dreams of being an athlete, not one of the guys in a tracksuit standing on the sideline. As a teenage goalkeeper, he came close to making the Dutch junior national team but gave up the pursuit after discovering he was prone to chronic bursitis of the hips. He stuck around just long enough to wonder why he and the other goalkeepers were expected to do the same fitness work as midfielders, who routinely do three times as much running during a 90-minute match. For that matter, he thought, why should midfielders, who spend most of the game at a jog, train like strikers, who are usually either walking or sprinting?

So Verheijen and his swollen bursas stayed in school, earning degrees in exercise physiology and sports psychology, and began moving up the coaching ranks. As he worked with teams at all levels, from the lower Dutch professional leagues to the South Korean and Russian national teams, Verheijen saw more that didn't make sense to him. Head coaches would put their players through the grinder in preseason training camps, making them practice morning and afternoon, with few days off. Training sessions often consisted of so-called small-sided games, in which teams of three or five scrimmage against each other. While good for skills practice, since each player touches the ball more often, small-sided games are more strenuous than real ones for the same reason — there's never a break in the action. When a player would get hurt in training or early in the season, coaches typically blamed it on a lack of sufficient fitness.

That didn't square with what Verheijen saw at all. Most players had more than enough strength and stamina to do their jobs. What they lacked was *freshness*. They were starting the season as tired as if they'd just come back from the World Cup. That's not even a figure of speech; the best soccer players do spend their summers and other breaks training and playing with their countries' national teams, just as the best club teams play extra matches throughout the season in supercompetitions like the Champions League. They then return from their "vacations" having rested less than everyone else. (When Ramsey pulled that hamstring, he had spent much of the preceding summer helping Wales reach the semifinal of the European championship.) Worst of all, in Verheijen's objective view, was the handling of players returning from injury. Because their time off had left them detrained — coach speak for "out of shape" — they were given extra fitness work to do on top of their usual load to catch up with the rest of the team. That resulted in them being even more exhausted, and more often than not getting injured again. "It's absolutely shocking," he says. "If I'm a player, I would take these people to court."

In theory, soccer injuries should be relatively manageable. Collisions happen, obviously, but they're not a central part of the game as they are in football, hockey, or rugby. The vast majority of injuries are noncontact, and it's an article of faith in sports medicine that noncontact soft-tissue injuries are preventable. Any activity performed at high-enough volume will take a toll on the body, but as risk factors go, soccer gets a relatively clean bill of health. The action takes place on forgiving grass, not joint-pulverizing asphalt or hardwood. The players have moderately proportioned bodies, not extremely fat, skinny, or muscled up. They move through a wide and natural range of motion; they're not stressing the same set of joints and ligaments and muscles in the exact same way over and over again, as a baseball pitcher or a swimmer does. In sports like baseball, tennis, and fencing, athletes can build up large strength-asymmetries between dominant and nondominant side, and asymmetries of greater than 5 or 10 percent are associated with a higher rate of injuries. In soccer, however, ambidextrousness is the ideal, if not usually the reality.

Yet anyone who watches soccer knows injuries — real ones, not the ones pantomimed while rolling around on the ground for a referee's benefit — happen all the time. A seven-year study of players in Europe's elite UEFA league found the average player suffers two per season, with the average length of absence being 18 days. That increases to 24 days for players suffering a recurrence of the same problem. Over the course of a full season, a team with 25 players can expect to experience eight or nine severe injuries lasting more than a month in duration. Just as NFL players dread a torn ACL above all things and hockey players live in fear of pulled groins, soccer has its defining injury: the hamstring strain. It accounts for almost 12 percent of all injuries in the sport. Another study, of Icelandic players, found the risk of hamstring strains goes up 40 percent a year with every year after a player turns 28. Within games, injuries of both the contact and noncontact variety get more frequent toward the end of each half as players tire out.

In 2015 the average Premier League soccer player earned more than $2 million per season, or $57,000 per week, with the best players making many times that. No owner wants to pay a player millions to sit on the bench; no player wants to lose out on his next contract because of a balky hamstring. The universal anxiety to avoid the injury bug has given rise to some odd practices. Diego Costa, Robin van Persie, and Frank Lampard are among the marquee footballers said to have received infusions of horse placenta from a Serbian therapist, Marijana Kovacevic, who claims it speeds healing. (Kovacevic's website is coy about exactly what's in the "original gel" she administers but promises, "All substances used during the treatment are completely natural and are safe for any type of anti-doping test.") For years, equally strange rumors swirled around what was going on at the football club AC Milan. In the early 2000s the club had a run of successful years, including winning the 2003 European Cup, with a lineup consisting heavily of players in their 30s and early 40s like Paolo Maldini and David Beckham. Media accounts invariably pointed to the work of the team's medical director, a Belgian chiropractor named Jean-Pierre Meersseman, who operated out of a facility he called Milan Lab. Meersseman maintained a veil of secrecy around his work, but the details he let slip were intriguingly weird: he claimed to have cured midfielder Clarence Seedorf of chronic groin pain by removing his wisdom teeth. Meersseman also claimed he cut injuries at the club by 90 percent and reduced the use of medications by 92 percent using methods like these. "It's not accepted in evidence-based medicine but I don't give a damn about that," he told one reporter.

More prosaically, teams like Manchester United and Barcelona FC have spent tens of millions of dollars on high-tech sports medicine facilities and employ small armies of doctors, physiotherapists, soft-tissue therapists, nutritionists, and other specialists in a bid to stem the flood of torn hamstrings and other ailments. None of it seems to matter much. If anything, the top players, the ones with their own personal wellness teams to complement the clubs', often seem to spend the most time on the injury list. Just ask any Arsenal fan.

None of that is any coincidence, according to Verheijen. More often than not, he says, teams that think they're using sports science to keep their players healthy are doing the opposite, because they don't know what they're doing. They're attacking the symptoms while ignoring the cause. Players in the world's top soccer leagues don't spend so much time injured because they're not fit enough or because their medical treatment isn't up to snuff, Verheijen says. They spend so much time injured because they're doing the wrong kinds of fitness work, at the wrong times and in the wrong amounts, and it's leaving them weak, slow, and vulnerable.

The fundamental concept underlying all fitness training is an idea called progressive overload. It was formulated by Thomas DeLorme, an army physician who designed rehabilitation protocols for injured soldiers after World War II. When the muscles are subjected to stresses they're not used to, DeLorme observed, they respond by breaking down a little bit, then growing back a little stronger. If you space out the stresses properly over time and increase their amplitude gradually enough so the muscles are always being challenged just the right amount, you can stack the strength gains on top of each other. The practice of constructing a program of progressive overload to optimize for performance at a specific point in time is called periodization.

Thanks to the work of people like Dave Costill, we know a great deal about how periodization works in endurance athletes. We know, for example, that marathon runners perform best if they taper off their training a few weeks before the race. Doing shorter, faster runs allows their muscles to build up energy-producing glycogen stores and readapt themselves to the faster movement patterns needed for a late-race "kick."

But soccer and other team sports have little in common with endurance sports. When Meb ran marathons, he trained for months at gradually escalating volumes and intensities, culminating in peak fitness on race day. A soccer season lasts for many months. Every coach wants his team in something like peak fitness for the start of it, and from that point, injuries and the need to rest players between games make it a struggle just to maintain. A 1988 study of hockey in the journal *Sports Medicine* found the average hockey player gains some anaerobic capacity over the course of a season but shows no change in aerobic fitness and actually loses some muscle strength. The authors of that study recommended that teams adopt "specifically designed strength maintenance programmes" for use during the season. (Like, say, slap-shooting a medicine ball against the wall?)

Moreover, so-called stop-and-start sports like soccer, hockey, and basketball are completely different from endurance pursuits on a physiological level, and the players who thrive in each have different biological makeups. As we've seen, great distance runners have muscles made up predominantly of slow-twitch, or type I, muscle fibers. The average soccer player has something closer to a 50-50 mix of fast and slow fibers, and the most sought-after players — the forwards like Lionel Messi or Arjen Robben or Cristiano Ronaldo, who can blow by defenders with their sudden acceleration — likely have more fast than slow. Athletes with a high proportion of fast-twitch fibers are at greater risk for muscle sprains. (In fact, at the professional level, fast-twitch-dominant players are underrepresented because so many of them wash out young from their injuries, according to one study of Danish footballers.) They also take longer to recover from exertion, since fast-twitch fibers, which are optimized for anaerobic respiration, don't receive as much blood supply. Verheijen compares explosive athletes to cheetahs, which can chase their prey at top speed for only a minute at a time before they have to stop and rest for an hour. Cheetahs have muscles made up of 70 percent type IIx, or superfast-twitch, fibers.

If you take the most explosive players and train them as though they're endurance athletes, with frequent, high-volume workouts, they're going to respond very differently. Instead of accumulating more and more fitness, they're going to accumulate more and more fatigue in a grim mirror-image of the progressive-overload principle. "If you are extremely fit but also extremely tired, your performance will be shit," Verheijen says. And shit performance isn't the worst of it. In the extreme case, the accrual of fatigue can develop into overtraining syndrome, a dreaded condition in which an athlete's energy level craters and his sleep and immune function are compromised. A full recovery from overtraining syndrome can take months. Then there's injury risk. "When the body accumulates fatigue, one of the characteristics is your nervous system gets slower. The signal from your brain into your muscles travels slower," Verheijen explains. "When you are landing or turning, the brain has to send signals to the muscles around your ankle and your knee to stabilize the joints. If you're fatigued, the signal arrives later, maybe too late. You are landing or turning but the muscles around your joint haven't contracted yet. You have an unprotected knee or ankle. Then your ankle ligament might snap or your ACL might snap. Then everybody says, 'Oh, bad luck.' No, it's accumulation of fatigue."

To Verheijen, all this was obvious. It would have been obvious to everyone in the soccer world, he says, except that so many of the professionals responsible for players' health had learned sports science that was mostly derived from the study of nonsoccer players — those runners and cyclists who are so easy to study in the

lab. Coaches were conditioning players to compete in sports totally unlike the ones they actually had to play, measuring their fitness with tests of aerobic endurance and comparing their blood
markers with those of distance racers instead of athletes in other
stop-start sports. "These sports scientists don't have a clue about
football. They're watching a football game but they don't see what
they're looking at," he says. "The traditional approach is to go from
theory to practice. I went from practice to theory."

So he put his theories into practice. He oversaw the fitness conditioning of Russia's national team during its unlikely run to the
semifinals of the 2008 European championships and did the same
for South Korea's squad at the 2010 World Cup, the first time that
nation ever made it past group-stage play outside its own borders.
As head trainer at Manchester City FC in 2009, he helped reha-
bilitate the career of Craig Bellamy, a 12-year veteran who was
one of the Premier League's most dynamic scoring threats when
healthy but had never made it through even half a 38-match season without an injury. Under Verheijen's care, Bellamy, restricted
to one game a week, made it through his first full season and put
together a string of 43 injury-free games. (The streak came to an
end only after the manager at his next club, Cardiff, put him back
on twice-weekly duty; Bellamy promptly tore a hamstring.) The
regimen that Verheijen devised for Bellamy — including substantially reduced training loads during practices as well as the one-
game-per-week limit — became his standard recommendation for
highly explosive players and those over 30. For all of City's players,
he cut preseason training to five sessions a week; players used to
working out twice as often had to be reassured they were still getting the conditioning they needed. Premier League teams can have
as few as two or as many as four days between games; Verheijen
performed a data analysis to show the disparity in rest was having
an undue effect on match results and used it to lobby for Europe's
soccer leagues to alter their scheduling. He changed the format of

the standard shuttle-run "beep test" used to assess conditioning, building in short rest periods to simulate the breaks in play during a game. Most crucially, instead of telling athletes just back from injury they needed to do extra fitness work to get back into "game shape," he periodized their returns, having them do less than everyone else and gradually increasing the amount until they caught up. Everywhere he went, he says, soft-tissue injuries fell and older players reverted to the form of their younger days.

It's hard to know just how much of the credit Verheijen deserves for the results his teams achieved. As an assistant coach, he has never had full control over players' programs, and as a journeyman, he has rarely stayed in one place for more than a few months at a time. (At Manchester City, for instance, he was let go halfway through the season when Roberto Mancini replaced Mark Hughes as manager.) But a Swedish physician named Jan Ekstrand has conducted extensive research into injuries in elite soccer and concluded that managers do, in fact, have relatively stable injury rates over their careers, even when they switch teams, and teams have different injury rates under different managers. (The exception: most managers experience a rash of player injuries upon taking over a new team. Ekstrand believes this occurs because new managers, anxious to establish themselves, work players harder and the players, anxious to impress their new boss, push themselves harder still.) When Verheijen hammers Arsène Wenger on Twitter for being a stubborn dinosaur, he may be guilty of rudeness, but he's not wrong to point the finger at the top.

Still, all over the world, managers like Wenger continue to push their players to the breaking point and then show surprise when they do, in fact, break. It's partly a cultural thing: most coaches were once players themselves, with little formal education to draw from; all they knew was what they had been taught. It's partly an optics thing: if a coach loses players to injury, most fans will blame luck, but if he reduces training and loses games to other teams that trained more traditionally, most fans will blame the coach. And then there's the sports world's quasi-magical belief in the ability of experts to overcome the limits of the human body. The world's best athletes are the kind of people who can train and compete obsessively; they expect the specialists around them to do their jobs as well as they do their own. Both athletes and coaches, meanwhile, are prey to the psychological phenomenon of "commission bias," which favors active responses like assigning extra fitness work or surgical intervention over passive responses like simply letting a player rest. "The problem is worst in the Premier League because all these clubs have so much money to spend," Verheijen says. "Football in the U.K. is hijacked by sports science. All these teams spend millions on the sports science department and it's quantity over quality. What you always see is if clubs have little money to spend they are very efficient, very creative. If people have too much money to spend, they buy a car they don't need."

Not that things are much better in America, where "more is more" is practically the national motto. "In the U.S., in sports, people think you can solve everything with fitness. That's how they think. That's the culture," he says. He has particular scorn for the NFL, where coaches run players ragged in the preseason, and takes a perverse pleasure in watching the HBO show *Hard Knocks*, a documentary-style look at NFL training camps. "The level of periodization in the NFL is absolutely shocking," he says. "In the preseason, these players are absolutely hammered. So, one after the other, they snap their ACLs. Fucking hell. What a bunch of amateurs.

"And this is objective fact," he adds. "This is not my opinion. I hope you understand that."

But it hasn't all been for naught. European soccer may be slow to embrace Verheijen, but Verheijenism has taken root on the other side of the pond, in the NBA. Injuries in basketball are no less prevalent than in soccer. In the 2015–16 season, the average player lost more than 12 games to injury, according to the blog *In Street Clothes*, which tracks player health statistics. While some of those missed games were on account of sundry ailments including "bed-bugs" and "severe testicular trauma," according to official team re-cords, ankle and knee sprains were the most common cause for absence.

In one crucial sense, in fact, injuries are much more of a factor. A basketball team consists of 5 players on the floor at a time, and rosters are capped at 15. Compare that with the much bigger ros-ters in other sports: soccer teams carry 25 or more players to field an 11-man side; a hockey team's roster is 23 players, with 6 on the ice at a time; and the NFL, with its hyperspecialization by position and absurdly high attrition rate, allows teams 53 roster spots. An injury to a core player in basketball is thus mathematically more significant than in any other major sport.

That fact was basically academic, however, until the early 2000s, when one team, the San Antonio Spurs, started having remarkably good injury luck. (It may have started even farther back than that, but the league changed its procedure for injury reporting in 2005 to discourage gamesmanship, making it hard to compare between eras.) It would stay remarkably good even as the Spurs, year after year, fielded one of the NBA's oldest rosters. Among aficionados of so-called advanced metrics, who crunch numbers hoping to find an advantage in fantasy sports contests, it's conventional wisdom that a team that has abnormally good or bad health one season will usually revert to the mean the next year. The Spurs haven't. They've been below the league average for injuries in each of the past 10 seasons. No team has lost fewer player-hours over that pe-riod.

For that, Spurs fans can thank the team's longtime head coach, Gregg Popovich, who was doing exactly what Verheijen has spent 20 years trying and failing to persuade European soccer managers to do: manage players' exertions to avoid the buildup of fatigue and give them a break once in a while. When you put it that way, it sounds hard to argue with. In fact, it was highly controversial. While the wisdom of holding key players out of meaningless games late in the season to rest them for the playoffs was generally accepted among Popovich's peers, the notion of doing it throughout the season struck many as bizarre and unsportsmanlike. At times, Popovich seemed to be thumbing his nose at his critics, as when the Spurs listed then-36-year-old Tim Duncan on an injury report as "OLD" to justify sitting him. In 2012 NBA commissioner David Stern slapped the Spurs with a $250,000 fine after Popovich sent four players, including three starters, home early from a December road trip; Stern said the coach had ripped off fans in Miami by withholding the stars they'd paid to see. Popovich was undeterred. Having already won four championships doing it his way, he continued to prove that players play better when they play less. In the 2013–14 season, which culminated in a San Antonio championship, no Spurs player logged more than 30 minutes in a game.

Popovich's success didn't go unnoticed by one of his former players, Steve Kerr, who had won two championships with the Spurs (and three more with the Bulls) before retiring to become a TV commentator. When Kerr reentered the league as head coach of the Golden State Warriors in 2014, he brought his old coach's philosophy with him. Unlike the Spurs, Golden State didn't have a lot of older players; its stars were younger guys like Stephen Curry, Klay Thompson, and Draymond Green. But the team had recently been bought by a group of Silicon Valley venture capitalists who believed in using technology and quantitative analysis to find small advantages it could exploit.

Overseeing that effort during the Warriors' 2015–16 season, when the team won an NBA-record 73 regular-season games, was its head of physical performance and sports science, Lachlan Penfold. An Australian, Penfold's experience was mainly in working with Olympic track-and-field athletes as well as Australian-rules football and rugby union teams. He appreciated the value of strategically resting players from his days in Aussie football. "You'd play twelve games, have a bye week, and then play another ten games. The difference in energy players came back with after their bye week was remarkable," he says. But initially, applying his experience was a challenge. He was used to working in sports where athletes trained far more often than they competed, giving coaches a high degree of control over when they peaked. The dense NBA schedule left him little time to work with the players — little time for anything but games, travel, treatments, and recovery. It was permanent triage.

"In track and field, you have this four-month block of programming where you know exactly what you're going to do," he adds. "You can't do that when you're competing every second day. Things change. You have to be very reactive. If you're playing four games a week, the traditional periodization model goes out the window. You've got fifteen different individual periodization cycles going."

To stay on top of those cycles, the Warriors use technology from an Australian company called Catapult, which makes small sensor-packed devices that fit under an athlete's jersey. Accelerometers in the device generate data on a player's direction, speed, and volume of movement; a software dashboard helps coaches make sense of the data, spitting out analytics that reflect a player's "training load." A sudden decrease in training load often indicates a player has an undiagnosed injury or illness; a sudden increase is a warning of an injury yet to come. A number of studies have shown that sudden jumps in training volume or intensity lead to injuries. According to one study of Australian-rules football players, when an athlete's "acute load" in a given week exceeds his "chronic," or average, load by more than 50 percent, his risk more than doubles. A similar analysis, drawing on two years of data, found that doubling a player's acute load results in a five- to eightfold increase in noncontact injuries. "Large week-to-week changes in load (rapid increases in intensity, duration or frequency) have been shown to place the athlete at a significantly increased risk of injury," concluded the authors of an International Olympic Committee consensus paper in 2016. Penfold says Catapult data is especially useful in monitoring the training output of injured players to make sure they're not ramping up too quickly to get back on the court sooner.

Catapult also has changed how the Warriors train as a team. When Kerr was a player in the 1990s, "coaches just ran the crap out of us," he has said. But accelerometer data from his players, backed up by an overhead camera system called SportVu that captures action during games, persuaded Kerr that all that running had been a waste of energy. Almost 85 percent of what players do in games, it turned out, is backpedaling or lateral shuffling, so Warriors coaches altered their regimen to reflect that. Other coaches boast about how hard their players work; Kerr is more apt to boast that his don't do wind sprints.

In addition to Catapult, the Warriors wear heart-rate monitors under their jerseys at practice to measure how well they're recovering from one day to the next. An elevated resting heartbeat can show physical fatigue while a decrease in heart-rate variability — basically, how much the heart speeds up and slows down in response to stimuli — reflects a tired nervous system. "It's really about what the minimum amount of work is I can do to have an effect without creating fatigue," Penfold says.

The minimum amount of work: that's a powerful idea, and one that's been gaining currency among fitness consumers thanks to the popularity of *The 4-Hour Body*, a 2010 book by the productivity guru Tim Ferriss, and *The 7-Minute Workout*, an interval-training app designed by researchers at the Johnson & Johnson Human Performance Institute. Ferriss also helped popularize the phrase "minimum effective dose," a term borrowed from pharmacology by Nautilus inventor Arthur Jones to apply to exercise physiology. Top sports scientists often talk about the minimum effective dose of fitness training, but not for the same reasons millions of people have downloaded *The 7-Minute Workout*. They don't care about freeing up their athletes' time for other pursuits or sparing them the unpleasantness of a marathon gym session. What they do care about — increasingly, what they care about more than anything — is avoiding the accumulation of fatigue. When athletes build up fatigue faster than they can recover, they perform worse, get hurt more, and retire younger, period.

Whereas coaches once thought in terms of how much load their players could sustain without a breakdown, now the smart ones focus on how little training they can get by on without a decrease in performance. But there's a problem with that formulation: it doesn't square with how we like to think about elite athletes. We get a kick from thinking about Adrian Peterson running 300-meter sprints under the Texas sun or Carli Lloyd spending her Christmas morning doing push-ups and sit-ups. We love hearing Donald Driver say, "When you're sleeping, I'm working," or Jaromir Jagr announce, "Physically, I don't get tired." It gratifies our desire for moral order in sports: we want the longest careers, like championship trophies, to be enjoyed by athletes who want it the most. It inspires us, and inspiration is a big part of what those of us who aren't elite athletes crave from those who are. "No pain, no gain" may be a recipe for failure, but it looks good plastered on the wall of a gym. "Do just enough," not so much.

It also tallies with the messages we've been fed by sports apparel and beverage brands, which have a vested interest in promoting commission bias. In 2015 Nike released a new basketball shoe, the Kobe X Blackout. The name was a tribute to Kobe Bryant's "legendary workout routine," according to marketing copy, which quoted him as saying he sometimes goes at it in the gym until he loses consciousness.

By the time the shoe appeared, however, Bryant was 36 and had long dispensed with the blackout sessions, according to his trainer, Tim DiFrancesco. "Everyone always wants to talk about his famous 'blackout' workouts," DiFrancesco wrote on his blog. "What impressed me more is what I saw from him later in his career. He trained hard with ruthless consistency but always knew when to call it a day well before hitting exhaustion.

"Approach your workouts with a 'live to train another day' attitude instead of a 'train until you puke' attitude. The approach you need for enduring success is one based on consistency."

We're so invested in the myth that athletic longevity like Bryant's is something that can be earned through suffering, we refuse to see the evidence even when it's staring us in the face. Like Arsène Wenger clinging to his belief that only unfit players get injured, we selectively edit the details we pay attention to and fail to realize the athletes we celebrate for working longer and harder than anyone else are, in fact, the ones who were ahead of everyone else in grasping the need to avoid fatigue. Donald Driver's off-season plyometrics workouts may have been intense and nonstop, but they only lasted 40 minutes each, three times a week, and that was his workout regimen for six months out of every year for years. Carli Lloyd may be a calisthenics addict, but leading up to a major tournament like the World Cup, she carefully tapers down her volume to ensure her legs will be fresh for competition. Meb could suffer with the best of them, but where his discipline and experience proved most useful was in knowing when one more training run would be counterproductive. "It's a fine line, and you'd rather be a little bit below that line, a little bit undertrained," he said.

And Jaromir Jagr? It's hard to imagine a guy who once drank twenty cups of coffee a day and says "My body doesn't get tired" worrying about the minimum effective dose of anything. But that turns out to be exactly what he does. According to Tommy Powers, the Panthers strength coach, all those middle-of-the-night workouts are instead of the team's regularly scheduled training sessions, not in addition to them. The other Panthers do their gym work in the morning, but Jagr says his body, after decades of playing mostly night games, doesn't feel ready for hard exertion at that time of day. A longtime practitioner of mindfulness meditation, Jagr doesn't need a heart-rate-variability app to tell him when to back off. "He's got this internal system that tells him when to go and how to go," says Powers. "He says you have to be excited to be there in order to get the results you're looking for."

When we look at Jaromir Jagr, we see what we expect to see: an athlete who has defied the very long odds against being able to do what he does at his age because he is such a maniac about his work habits. In fact, he's someone who — in addition to being preternaturally huge and fast and coordinated — is wildly successful at knowing how much and when he can work without it being too much. "He has a balance other people might not consider a balance, but for him it's totally fulfilling," Powers says.

3

WRESTLING WITH GIANTS

The Surprising Benefits of Cross-Training

D on't think how to do it. Do it. Do it. Do it. Do it. Do it. Do it.
Do it.

"When you think you've done it enough, do it a thousand times more."

In a small, brightly lit martial arts gym on the far western edge of San Francisco, a short, powerfully built man with a black buzz cut is giving a lesson on the easiest way to break someone's arm. The man, Carlos "Sapão" Ban, is a former three-time jujitsu world champion. He learned the fighting sport at the hands of Ralph, Renzo, and Ryan Gracie, scions of the clan that essentially invented the style known as Brazilian jujitsu and introduced it to America, where Royce Gracie used it to utterly dominate opponents in the early days of the Ultimate Fighting Championship. Seated on the floor mat in a semicircle around him are his pupils, eight men and one woman, ranging in age from early 20s to mid-40s, clad in pajama-like *gis* of sturdy white cotton, cinched with belts of white, purple, or brown.

Alex Martins wears purple, a strip of cloth he's never washed, in keeping with jujitsu tradition. A competitive big-wave surfer and owner of a local surfboard shop, he has been coming here two to three times a week for the past five years, perhaps a little more or less often depending on how the waves are breaking on Ocean Beach, a few hundred yards away. Eight weeks ago, he competed in the invitation-only contest at Mavericks, a renowned offshore break half an hour down the coast from San Francisco. Mavericks is one of the marquee events in the small-money world of big-wave surfing, and the only one to be held in continental North America. It's held only when the forecast calls for just the right conditions, can take place anytime from November through March on a few days' notice, and some years it doesn't happen at all. If an irregular event can have regulars, Martins is one. All winter long, upon waking up, he checks webcams that show the buoys at Mavericks to see if the waves there are right for riding. On mornings when they're not, he rides his home breaks on Ocean Beach, itself a world-class surfing spot for those who can tolerate the chilly fog and frigid water.

But this year's surfing season has come and gone. It's April and the arctic winter storms that send monster swells marauding down the coast have lost their energy. The prevailing winds are now blowing from the south, turning orderly waves into a boil of incoherent chop. For the next few months, jujitsu, yoga, and physical conditioning will be Martins's life when he's not repairing surfboards or home with his wife and two young sons. Compact and sturdy like Sapão, with startling ice-blue eyes in a weathered face, Martins is, at 45, the oldest student in today's class. He's not the only surfer, though; one of the other guys is wearing what's clearly a rash guard under his *gi*. Martins watches intently as Sapão demonstrates, once again, how to maneuver one's grappling opponent into a submission hold called an arm bar, using his legs to control his sparring partner's torso and hyperextending his arm out away from his body so that it would only take a nudge to dislocate the elbow.

Once the students have practiced the arm bar for a while, Sapão shows them how to defend against it, spinning out of his assistant's attempt and enveloping him in a cross-body cradle hold. Then it's an even more complicated sequence that involves "shrimping," or curling up on one's side in a fetal position, to escape being pinned, and somehow getting from there into applying a chokehold with the lower legs. Trying to keep track of the moves and counter-moves is brain-twisting stuff, like solving a 3-D puzzle while simultaneously being a puzzle piece trying to avoid getting solved. At one point, Martins and his practice partner get tangled up and have to talk through who's supposed to be doing what, their bodies pressed together and their faces inches apart.

"It's all mechanics, no power," Sapão lectures. "You guys need to follow what you're seeing, not what you're thinking. If it doesn't make sense, that's good. That's what a beginner should feel. You guys have a lifetime to learn this, so don't worry."

Then it's on to the next portion of class, sparring. Sapão pairs the students for five-minute bouts, rotating them after each. Martins duels with the lone woman, whose *gi* rides up to reveal faded pregnancy stretch marks. He fights the biggest guy in the class, who must have half a foot on him, to a standstill. The matches all have a similar rhythm: long periods of slow grappling, with a lot of feeling-out and subtle positioning adjustments, alternating with a few seconds of intense struggle in which one fighter or the other attempts a finishing move. By the last bout, the air in the gym has grown humid and funky with sweat. Martins, breathing hard, is fighting a white belt; his eyes have a thousand-yard stare but his hands keep probing, trying to gain control of an unsecured arm or leg.

"I want to hear you breathe," Sapão calls out. "If you don't breathe, you're going to get tired fast and end up tanking. You don't want to be checkmated. You want to be the one that does the checkmate. Breathing is what's going to save you guys."

That one Martins doesn't have to be told.

Like the style of jujitsu he practices, Alex Martins is from Brazil (his surname is pronounced mar-TEENS). Born in Recife, a tropical city 500 miles from the equator, he competed on the professional surfing circuit as a teenager, but the lack of economic opportunity at home made it hard to sustain the lifestyle. At 23 he emigrated to California, where, unable to speak much English, he made a living washing dishes and doing pizza deliveries. For the first few years, he had neither the time nor the money to resume surfing, and his late restaurant hours made it hard to join the "Dawn Patrol." (Dedicated surfers are up early because later in the day, air rising off the warming land creates a low-pressure zone that pulls in cooler air from out at sea. The resulting onshore breeze collapses waves.)

By 2001, however, Martins was making enough to stay ahead of his bills and had started surfing again. The cold, powerful Northern California waves were nothing like the ones he'd grown up with. It wasn't long before friends asked if he wanted to come along to surf at Mavericks. He didn't. He knew the spot's reputation as one of the world's most dangerous breaks. People have died at Mavericks, and not just any people, but expert big-wave surfers, the kind who spend the year hopscotching around the globe chasing swells. In 2011 a Hawaiian named Sion Milosky, famed for having ridden the biggest wave any surfer had ever caught without a motorized tow-in, drowned after wiping out at the end of a long afternoon of surfing 50-foot ledges. He followed another Hawaiian pro, Mark Foo, who became the first surfer to drown there in December 1994.

Surfing anywhere is a risky proposition, with hazards that range from sharks to razor-sharp coral reefs to getting tangled up in one's own ankle leash. Surfers get plowed into the sandy bottom and break their necks, or collide with their boards in the foamy roil of a wave's impact zone. But Mavericks is uniquely treacherous. The underwater topography at the break, which sits a half mile offshore, accelerates the outside edges of the wave faster than the middle, giving it a V shape with a bowl in the middle. A surfer who fails to make it out of the bowl experiences the equivalent of a tsunami concentrated into a single point. As if that weren't enough, underneath the convergence point is the Cauldron, a hole in the ocean floor that acts like a suction cup as the waves recede, drawing water, and surfers, to the bottom.

Martins knew all this and wanted no part of it. He's not an adrenaline junkie but a hardworking middle-aged dad whose bedtime is 9:00 p.m. It's the challenge that draws him to surfing, not the danger. When his friends invited him to surf Mavericks, he agreed to accompany them but resolved not to attempt the main break. Even though the waves were small by Mavericks standards — "only" 20 feet, not the 40 or 50 feet they can often reach — he stuck to the shoulder of the wave and watched his friends attempt its peak. The longer he watched, though, the more he found himself intrigued — not so much by the thought of surfing Mavericks as by the thought of becoming the kind of surfer who belonged there.

The problem was how to do that. Mavericks is an order of magnitude more fearsome than any other break in California. There was no Mavericks Lite he could frequent to train for the real thing, no training wheels version. "For me, to go from surfing Ocean Beach to surfing Mavericks, it was like, 'This is crazy,'" Martins says. "You'd see guys taking off and dropping into a wave, and think, 'Oh my God. Hopefully he's not going to die.' In my mind, there was no way I could survive one of those wipeouts. I did not see myself as prepared for that, either physically or mentally."

Martins had read that Jeff Clark, the first surfer to successfully tackle the main break at Mavericks in 1975, was a proponent of yoga as a way to build flexibility and core strength, so Martins started taking classes in an athletic form called ashtanga. The mindfulness aspect of yoga practice also proved beneficial; when navigating huge surf, he says, letting your focus wander even for a moment can be fatal. Martins earned a teaching certification in ashtanga but stopped giving lessons after opening his shop in 2010. He also started working out three times a week with Simon Fathers, a performance coach from New Zealand who offers a conditioning program tailored to surfers, heavy on rotational whole-body movements that demand balance and efficient power transfer between core and extremities.

"If you want to do big waves, you can't just surf," Martins says. "Everything I do, I started because I wanted to get fit and surf that place."

As Martins gained notice on the local surfing scene, he received his first invitations to the contest at Mavericks, which Jeff Clark and others had been running since 1999. He was named as an alternate in 2003, and then as one of the two dozen invited competitors in 2006. In 2010 he made the semifinal round at what was believed to be the first-ever paddle-in surf contest held on 50-foot waves. Three years later, he achieved his best finish ever, coming in fourth in the finals. By then, Martins was 42. In most sports, that's a late age to be peaking, and most sports don't require surfing's combination of high fitness and higher tolerance for risk.

But big-wave surfing is a curious exception to the usual order. Older athletes don't just hold their own; they fairly dominate the sport. Fully half the riders at the 2016 installment of Mavericks were in their 40s, and five were older than Martins. Economics has a lot to do with that. The vast majority of the prize and sponsorship money in surfing goes to riders who compete on smaller waves using short boards. Compared with the giants at Mavericks, which demand every ounce of a rider's craft just to stay upright, small waves allow for more intricate, spectator-friendly demonstrations of skill, and they're also easier to schedule events around, since smaller breaks are more numerous and reliable. Big-wave surfing has thus always been a niche sport. The people who do it tend to be, as psychologists would say, intrinsically motivated, and even the elite ones have day jobs. Martins's Mavericks competitors this year included a filmmaker, a landscaper, a magazine publisher, two paramedics, and a geologist.

That might be changing. Money is starting to make its way into big-wave surfing. In 2014 a sports marketing agency called Cartel Management bought the contest and fronted a $120,000 purse. (Three years later, Cartel would declare bankruptcy, leaving the future of the contest in limbo.) At this year's event, a 26-year-old named Nic Lamb took first place and won the champion's prize of $30,000. Martins had the bad luck to share a midround heat with Lamb and the two runners-up, both in their early 30s, and accepted an early exit on a day when mediocre surf made it hard for the true big-wave specialists to distinguish themselves. He expects the competition at Mavericks to get younger as marketers wake up to big-wave's appeal and the financial rewards attract entrants from the youth-filled pro circuit.

But the small purses aren't the only reason older guys have ruled Mavericks. In Martins's view, tackling the biggest, trickiest waves requires a level of experience and discipline most surfers in their 20s just haven't developed yet. Like many other extreme sports, big-wave surfing only seems like it's for daredevils. In reality, the ones who excel over long periods of time are, like Martins, highly calculated in their risk-taking and methodical in their preparation. Reckless types don't last long. There's a saying about aviators — "There are old pilots and there are bold pilots, but there are no old, bold pilots" — that applies equally well to big-wave surfers, with their similarly obsessive attention to equipment and weather.

Even recreational surfers are notorious for organizing their lives around their hobby, keeping wetsuits in the car and planning vacations around the swells. Martins takes it to another level. He credits his longevity to the yoga and the strength work as well as to his early bedtimes and clean eating habits. It's an entire lifestyle he could only imagine when he was new to San Francisco, scarfing free pizza for dinner at midnight every night. "Even though I'm not a professional surfer, I live my life as if I was," he says.

Even with all that in place, Martins has never lost his fear. Every season at Mavericks, he says, "there are going to be three or four days that are gonna scare you, days that are like, 'Whoa, what am I doing here?'" Wipeouts are inevitable and unpleasant everywhere, but at Mavericks, even with modern safety measures like Jet Ski patrols and inflatable safety collars, they're singularly terrifying. "Imagine a wave as big as a big building. The force of the water is like — it's hard to explain," Martins says. When he falls, the wave drives him down so deep, he doesn't even bother to open his eyes because there's no light to see by. "You can easily die here just because the wave is so violent."

In 2014, tuning up at Mavericks three days before the contest, Martins paddled into what looked like a 30-foot wave but realized immediately upon popping up that it was moving too fast for him to cross the face and make it onto the shoulder before the curling wall of water mounting over his head collapsed onto him. He tried to outrace it by pointing his board down the slope but accidentally went airborne as he dropped off the overhang. Skidding out of his "air drop," he overbalanced and toppled off the nose of his board. He hit the water so hard he skipped off it like a flat stone, then looked up to see the wave he'd been hoping to outrun breaking onto his head. The impact plowed him deep under the surface and kept pummeling him deeper, what surfers call "the pound." When this happens, they're taught not to resist but to stay calm until the wave has passed and only then kick to the surface. The calmer you stay, the less oxygen you use. "Fighting, that's what's going to kill you," Martins says. "It's not time to freak out, because the wave's not going to release you at that moment. You have to relax and take the pound. But that wave, I felt like the pound would never end." By the time it did, the wave had driven Martins so far down, he couldn't make it up to the surface before the one behind it was breaking on him. "That's what we call a two-wave hold-down," he says. Two-wave hold-downs are what surfers fear most; that's exactly how Sion Milosky is believed to have drowned. Since waves usually come in sets of four or five, Martins often declines the first wave or two of a set to reduce his risk. Now he wasn't sure he could outlast even one more. "That was the only time I felt like I was going to die," he says.

Fortunately, the next wave wasn't as powerful, but as Martins kicked toward the light, he realized his knee, wrenched this way and that as he'd hit the water and been tumbled underneath it, was badly damaged. Back on land, an MRI revealed he'd torn his ACL, MCL (medial collateral ligament), and meniscus. A surgeon took a graft from his patellar tendon and sutured it in where his ACL had been. It was a year before he was ready to surf again.

That experience, along with other, less dramatic wipeouts, is why Martins is here at jujitsu class. As with yoga, he took it up after hearing other surfers praise its benefits. But whereas yoga helps him perform better on the board, jujitsu is all about what happens when he falls off it. Getting pummeled in the bowl at Mavericks, he says, is almost exactly like fighting a much bigger, stronger opponent who's using his weight to mash you into the floor. In both situations, the untrained reflex is to panic and struggle. In both situations, that's the worst thing you can do, because there are only two ways things can go bad in a hurry. First, you can lose control of one of your limbs. Do that in a grappling bout and you'll fall prey to an arm bar or leg lock; do it in the Mavericks washing machine and you could end up in the hospital like Martins. (At the moment his ACL blew, "it felt like someone grabbed my leg and went like this," he says, making a wrenching motion.) Second, you could run out of oxygen. Just as a held-down surfer must slow his heartbeat and wait for a lull between waves, a pinned wrestler's challenge is to conserve as much energy as possible while he waits for an opening to attack. The wrestler has to remind himself to breathe and the surfer has to fight that reflex, but they're really doing the same thing: avoiding oxygen debt.

Some of the people Sapão trains are competitive mixed martial artists. Martins tells them he thinks they must be crazy to get in a cage with another fighter looking to maim them. "And they say, '*You're* crazy. If I'm getting destroyed, the referee will stop the match. There's no referee inside the ocean.'"

As we've seen, when elite athletes continue to succeed well past the peak age for their sports, it's not because they train more but because they train more efficiently. That means they use periodization to get the conditioning and skills practice they need without accumulating unnecessary fatigue, especially in sports where conventional periodization is difficult. But for athletes seeking world-class status, balancing the imperatives of fitness and freshness is just the start. They need to locate and exploit every small advantage. For the older athlete, that means finding innovative modes of training that allow them to do as much effective work as possible without subjecting their bodies to extra wear and tear. That's where cross-training comes in.

Alex Martins's jujitsu regimen is an extreme example of an athlete using cross-training to maximize his longevity. It allows him to rehearse for the most dangerous part of his job without exposing himself to 50-foot waves, sharks, undertow, underwater boulders, or any of the other things that make big-wave surfing so dangerous. Usually the benefits of cross-training are more prosaic: athletes do it to improve different aspects of their fitness, hone new movement patterns, build up muscles that otherwise get neglected, and above all add volume to their workouts without subjecting their bodies to more of the same repetitive pounding, twisting, and grinding they get in their day jobs.

It can also yield useful new skills. Steve Nash, who was named the NBA's MVP twice and retired at 41, reserved his off-seasons for soccer, his first love. At his playmaking peak, Nash saw passing lanes no other point guard would have because he was sometimes literally playing a different game; occasionally he even busted out an actual soccer trick on the court. Terence Newman, the standout Minnesota Vikings cornerback who started the 2016 season as the oldest player in the league at his position by five years, eschews his team's off-season training program, staying in shape by playing pickup basketball games; the footwork required to play defense on a basketball court, he says, translates directly to the task of covering wide receivers. The downhill skier Bode Miller taught himself how to walk on a tightrope-like slack line, claiming it helped him maintain his balance while slaloming around gates at 85 miles per hour. Quarterback Drew Brees practices staying upright by stand-up paddleboarding in the Pacific near his home in San Diego.

Cross-training isn't just for older athletes, of course. In fact, in the last few years there's been a major movement in youth sports to discourage early specialization and year-round participation in one sport. Led by Dr. James Andrews, a 75-year-old orthopedist and probably the only sports surgeon the average ESPN viewer can name — he has operated on Brees, Jack Nicklaus, and Michael Jordan, among others — the campaign is a response to a trend of the last few decades in which parents encourage their children to commit to a single sport early in hopes of securing a valuable college scholarship while club teams or academies take the place of vanishing physical education classes and school athletics. That trend has resulted in an epidemic of overuse injuries in adolescents that were once the province of adults: Little League pitchers with torn ulnar collateral ligaments (UCLs), middle school midfielders needing new ACLs. And those are only the short-term health impacts, says Nirav Pandya, an orthopedic surgeon at UCSF who works with both youth and adult athletes. "What we're finding out is these kids, transitioning to adults, they're so burnt out on sports that they don't want to work out when they're older," he says. "You're creating this culture where sports are a means to an end rather than just going out and playing. We should be encouraging kids at a young age just to be healthy."

But there's a reason for early specialization. It almost always requires thousands of hours of focused practice to develop elite-level skills, the kind you need, say, to win a full-ride scholarship or get drafted straight out of high school. Drawing from the research of psychologist K. Anders Ericsson, the pop-sociology writer Malcolm Gladwell popularized this idea in his book *Outliers* as "the 10,000-hour rule," although Ericsson himself says there's no hard and fast formula for translating practice time into expertise. In any case, it's fair to say it takes years of practice to become world-class at something. Once you're there, however, it takes considerably less practice to stay there. Ericsson calls the latter, more time-efficient work "maintenance practice." In a 1996 paper, he and a collaborator, Ralf Krampe, reported that older expert pianists exhibited a performance level comparable to their younger counterparts while putting in fewer than half as many weekly hours at the keyboard — 10.8 hours for the older ones versus 26.7 for the youngsters. Pianists make for a simplified model of skill retention over time since aging-related physical decline doesn't affect their performance the way it does athletes. But a sports psychologist named Janet Starkes looked at masters track-and-field athletes and came away with much the same conclusion, finding they were able to maintain their age-adjusted performance level on 6.5 hours a week of training, versus the 20-plus hours they had put in when young. Partly that's because it takes less time to rehearse skills than it does to learn new ones. But it's also, Starkes found, because masters athletes use their training time more efficiently. Whether expert or novice, young athletes, she wrote, "typically opt for the comfort zone that the rehearsal of well-versed skills affords." In other words, they waste a lot of time showing off what they're already good at. Masters athletes, in contrast, often focus their workouts on "elements needing remediation" and show more willingness to "expend . . . the effort to acquire or refine unstable skills."

That's about how it worked for Oksana Chusovitina, the 41-year-old gymnast from Uzbekistan who made it to the vault finals at the 2016 Summer Olympics while training 15 to 18 hours a week. Pretty much all of her competitors were within a few years of 18 and trained twice that amount. Chusovitina told the *New York Times* that "thinking like an adult" had made it easier, not harder, to compete with age. "I'm smarter," she said. "I know how to push myself." And 15 hours is an eternity in the gym compared with the regimen of Jordan Jovtchev, a Bulgarian who made it into the rings final at the London games while limiting his practice to 90 minutes a *week;* the 39-year-old said that was all his surgically repaired shoulders could tolerate. ("Gymnastics is a painful sport," he told the *Times*.)

Even athletes who don't practice at all retain some measure of the benefits from the intensive training of their youth, the thousands and thousands of hours spent grooving specific neuromuscular patterns into their synapses. When David Costill brought his runners back to the Human Performance Lab and retested them after 25 years, he was surprised to see what happened when he put the ones who had stopped training altogether onto the treadmill. Even though they had to work at more than 90 percent of their cardiovascular capacity, huffing and gasping just to sustain a modest eight-minute-mile pace, their biomechanics still looked like those of the elite performers they had been decades ago: "While these runners were chubby and out of shape, you could tell by watching them run on the treadmill that they had a gift for running."

Being able to practice less — being able to merely rehearse skills rather than do the time-consuming work of acquiring them — is a big advantage older athletes have that rarely gets talked about. It's an advantage because it lets them use all those extra hours doing other parts of their jobs — getting recovery treatments, or working on their mobility, or studying game film. The cycling coach Sir David Brailsford calls all the extra things athletes do "one-percenters," meaning that's how much difference they make in performance. Brailsford's teams have won 18 Olympic gold medals and four Tour de France victories in following an approach he calls "the aggregation of marginal gains," which covers everything from riders' posture on the bike to the bedding they use on race nights to how they wash their hands. (As well as, allegedly, the use of banned substances.)

Supplementing regular training with cross-training can be a big source of marginal gains. Bob Larsen believes more than anything that it was Meb Keflezighi's devotion to what he calls "ancillary training" that kept him at the top of American running into his 40s. "A lot of guys and gals will put the miles in. If they knew for sure that 200 miles a week would make them a champion, they would run the 200 miles a week," Larsen told me. "But they may not do these extra things because they would not be totally convinced how effective it might be for them. That's kind of the missing link for a lot of them. Meb, if you mention one thing that may be good, he's going to do it and he's going to continue to do it indefinitely." Four or five times a week, Meb augmented his training runs with a session on an ElliptiGO, an outdoor bike that operates like an elliptical machine. He also practiced aqua running, which is basically just jogging in a pool. Meb was one of the relatively few top marathoners who made cross-training a substantial part of his workout even when he was healthy, but the cross-training workouts came in especially handy when dealing with an injury that prevented him from running for a while. "He'll train really hard, just like he would if he's on the track, and then, within a couple weeks of him getting out of the pool, he's running very fast because he's very fit," Larsen said of Meb's practice.

The surfer Laird Hamilton takes the concept of aqua running — adding water to an exercise to reduce the impact and crank up the resistance — to another level with a workout he developed and patented called XPT, for extreme pool training. Inspired by ancient Hawaiian strongmen who would compete at lugging boulders along the ocean floor, Hamilton works out with dumbbells at the bottom of a swimming pool at his homes in Malibu and Hanalei. One of the few big-wave surfers to attain real fame beyond his sport, he came up with XPT as a way to maintain his hulking physique after a lifetime of injuries that have included five broken ankles and something like 1,000 stitches. "We're so trained to believe that unless we feel like we've been beat up, we don't think we got a workout," he told me. "At the end of the day, you can work harder than you've ever worked and come out and not feel beat up. That's the kind of fitness that, as I go, I like more and more."

At 52, Hamilton still surfs, but not competitively anymore. Instead, he has focused his energy on other board sports — stand-up paddleboarding, windsurfing, kitesurfing. "A big part of it is just enthusiasm," he says. "I keep creating this variety within the same arena and that can continue to excite me.

"I've always loved the term 'victory through attrition,'" he adds. "When you're the last guy standing, you don't even have to be any good. You just naturally win."

If anyone understands the victory-through-attrition thing, it's Hamilton's fellow big-wave surfer Alex Martins. As a 23-year-old in Brazil, he was a good surfer but not among the best around; had he been, he probably would have stayed pro instead of moving to San Francisco to wash dishes and learn English. When you're young, he says, success in sports is all about inborn talent — who has it and who doesn't. Over the years, however, the people who rely on talent alone drop by the wayside, either because they don't keep themselves in shape and get injured, or because they're unwilling to do even the minimum amount of work required of them, or because they become irrelevant when less gifted but harderworking competitors finally catch up to them. "Sometimes it evens out," Martins says. "I don't consider myself a supertalented guy. I consider myself a very hardworking guy. For me, the reason I'm still doing this is I feel like I've worked really hard for it."

Having worked that hard — hard enough that he was able to enjoy his best day of surfing ever at 42 — he's not ready to quit, and he doesn't have to. "If you're wondering why at age forty-five I'm still doing this, to be honest, right now I'm feeling in my best shape. I wouldn't have been able to do anything like this in my twenties.

"People say, when are you going to stop surfing Mavericks? My answer is, I think Mavericks is going to tell me when I should stop."

4

TRICK THE MIND, TRAIN THE BODY

How Technology Makes
Workouts Smarter

n a high-tech performance lab in Victoria, British Columbia, a petite, fine-featured 34-year-old woman is running on a strange-looking treadmill. From the waist down, she and the treadmill are both encased in a transparent plastic bubble. The woman, Hilary Stellingwerff, is one of Canada's best middle-distance runners. Five months from this March morning, she'll represent her country at the Olympics for the second and last time. Although she won't medal, just getting there will prove to be an important victory for women athletes throughout Canada. After giving birth to her first child in 2015, Stellingwerff took a year and a half off to be a full-time parent. Upon her planned return, the agency that disburses funding for Canada's Olympic hopefuls declined to renew her stipend, saying its rules allowed only 12 months for recovery from an injury. Stellingwerff took the agency to court to argue that treating a pregnancy as an injury represented discrimination against women, and won.

Now she's back on the periodization timetable her coaches have mapped out to ramp her up to world-class speed in time for Rio. A couple of months ago, however, around Christmas, she tweaked something in her knee that required several weeks to heal. Her primary coach, Dave Scott-Thomas, is located in Guelph, Ontario, 2,600 miles away, and the coach who oversees her training locally has reason to be extra cautious about her health: he's her husband, Trent Stellingwerff, the head of innovation, research, and physiology for the Canadian Sport Institute's Pacific branch and one of Canada's foremost sports scientists. A former high school track star himself, Trent loves his job possibly more than anyone I've ever met. His outsize friendliness and penchant for malapropisms soften his habit of saying things like, "If you remember from high school chemistry, carbon usually has a molecular weight of twelve."

Hilary has already done some road work this morning, jogging two miles from their suburban home to CSI/Pacific, but she has a five-kilometer road race coming up in two days and Trent doesn't want her to overdo it. "As you know, she's an older athlete so we need to be a little more intelligent with her total loading," he says. "And so we offset some of the running with work on an antigravity treadmill."

That's the nature of this weird-looking $75,000 contraption, made by a California company named AlterG. Trent demonstrates how it works. Hilary is zipped into the inflatable bubble by a rubberized disk she wears around her waist, much like a kayaker's waterproof skirt. "Right now, she's at ten percent deload," he says. Air pressure inside the bubble is buoying her up just enough so that she's striking the treadmill as though she weighed 94.5 pounds rather than her real 105. Trent presses a button on a control panel; a whoosh of air ensues and the bubble creaks. "Now we've deloaded her to eighty percent," he says.

"Ah!" Hilary cries as the skirt lifts her up that much more. "It's like running in space. I've never been able to handle much less than seventy-five without feeling really strange."

Middle-distance runners don't get much attention from the outside world, at least not in North America. The 100-meter race is the marquee track-and-field event of the Olympics, while the New York City and Boston Marathons draw millions of spectators. There's no in-between equivalent. But ask a group of runners what they think is the toughest event and most of them will tell you the 400, 800, or — Stellingwerff's specialty — the 1,500. All three combine the agony of a marathon with the intensity of a sprint. That's a function of physiology. Marathoners rely largely on their aerobic fitness and sprinters on their anaerobic capacity, but middle-distance runners are hammering both energy systems almost from the beginning of a race to its end. They suffer like I did in my encounter with the VersaClimber, without any nonsuffering part of their brains or bodies to retreat into. When Stellingwerff runs the 1,500, she's operating at 115 to 120 percent of her VO_2 max, consuming all the oxygen her blood can send to her muscles and going deeply into oxygen debt. A race lasts about four minutes, and she spends much of that time feeling "burning, nauseous, like you're gonna puke, that kind of thing, and you have to push through that," she says matter-of-factly. That she can do this for a living, and enjoys it, is only because she's "psychologically diabolical," in the words of her admiring husband, a miler himself in his competitive running days.

As you might guess, middle-distance runners typically achieve their career bests later than sprinters but before long-distance runners, putting Hilary on the outer edge of the age envelope for her event. From a neuromuscular standpoint as well, middle-distance running sits somewhere between sprinting and long distance. If elite marathoners have mostly slow-twitch muscle fibers and elite sprinters mostly fast-twitch, the muscles of a great 1,500 performer like Stellingwerff are usually full of an intermediate variety, so-called type IIa fibers. Stellingwerff wears spikes similar to a sprinter's when she runs on the track and has an average ground contact time of 120 milliseconds, not much more than a sprinter's, while taking 205 to 210 steps per minute. The more she can sustain the biomechanics of a sprinter throughout the interminable four minutes of one of her races, the faster she'll run. That's where the AlterG helps.

"Up until now, if you had an injured runner, you'd do pool running or you'd do cycling or the gerbil machine, the elliptical," Trent says. "But if you look at how slow the motor patterns are doing those things, it's not like what you see here." You can't take 205 steps per minute in a pool or practice *pop-pop-popping* off the ground with power on an elliptical. That's crucial because whenever you're training your body for something, you're always detraining it for something else. When a runner like Hilary does a high volume of running at a lower cyclic frequency, she's reinforcing the neural pathways that help her operate at that frequency and weakening the pathways that would help her maintain a higher cadence. That might be a worthwhile trade-off for a marathoner like Meb, whose success is mostly a function of his aerobic conditioning. But for someone who spends her entire race at a high cadence, a tiny decrease in foot turnover is a big deal. On the AlterG, that doesn't happen. Ground-contact time remains essentially the same; instead, her stride length increases to compensate for the decreased resistance. Hilary runs around 100,000 steps in a typical week of training, striking the ground with a force of four times her 50-kilogram body mass on each step. A 10 percent reduction in load is thus 40,000 kilograms' worth of force her bones, ligaments, cartilage, and tendons don't have to absorb. "So this offers a highly, highly specific mode of cross-training that doesn't mess up her nervous system and that really does mimic running well, that neural input and cyclic frequency," Trent says. "To try to mimic that in a pool or on a bike, forget it. It's too slow."

That's not to say all of Stellingwerff's training is speed work. Quite the opposite. About 80 percent of the running she does in a typical week is done at far below race pace, at the comparatively languid clip of a fit amateur. Stellingwerff's training program is structured around a newish but fast-spreading concept in endurance sports called polarization. Rooted primarily in the work of a Norwegian sports scientist named Stephen Seiler, polarized training aims to promote two different types of adaptations in muscle fibers: long, slow workouts drive an increase in the number of mitochondria (which, again, are the parts of cells that convert carbohydrates into usable energy), and short, intense intervals cause those mitochondria to increase their peak output. Studies have shown polarized training yields faster gains in VO_2 max, peak power output, and time to exhaustion than training plans focused on high-intensity intervals or lactate-threshold workouts.

Half of this equation might sound more familiar than the other half. The last few years have ushered in a much-deserved vogue for intensity in the fitness world, with mountains of new research showing the surprising benefits of short-duration, high-effort exercise, especially for older or more sedentary adults who don't regularly get that kind of exercise. To reap the benefits of intensity, you don't have to do very much work at all. One 2014 study found overweight adults who did a single minute of high-effort cycling three times a week (plus nine minutes of warm-up and cool-down, for 30 minutes of total weekly exercise) increased their cardiovascular fitness by 12 percent and saw a significant drop in blood pressure. But you do have to work out extremely hard during those short intervals: at intensities of less than 80 percent of VO_2 max the positive effects quickly taper off. Going for two minutes at 40 percent doesn't have the same effect; there's something particular about redlining the system that spurs hyperadaptation, even in tiny doses. In magazine articles and TV morning-show segments, in books like *The 4-Hour Body* and apps like *The 7-Minute Workout*, intensity has been portrayed as a magic bullet, and fitness franchises that offer some form of high-intensity interval training (HIIT), like CrossFit, Orangetheory, and Barry's Boot-camp, have been among the industry's fastest growing.

HIIT is an amazingly time-efficient way to build power and endurance and ramp up your metabolism, and advocates say it's especially valuable to older athletes. In the book *Fast After 50*, 61-year-old cyclist Ned Overend tells the author Joel Friel the secret to winning races in his 50s was shorter, harder rides: "I've learned that by reducing volume, I'm more rested for high-intensity sessions, and by being rested I can push myself harder during the intervals." Asked how runners can stay fast in middle age, Mayo Clinic researcher Michael Joyner told the *New Yorker*, "There are only two 'secrets.' Keep your VO_2 max up by doing intervals, and don't get injured."

That's all well and good for the average fitness enthusiast; unlike plenty of other health and wellness fads, HIIT is one that actually works. But for an Olympic middle-distance runner — or for a stop-start athlete whose games can last hours and involve lots of sprinting — it can't be a question of either-or. Volume is important, too. And intensity comes at a high cost. "As a coach you're always dealing with three inputs," explains Trent: "What's the stimulus and adaptation I'm looking for out of this workout? What's the recovery profile gonna look like? And what's the residual? That is, how often do I need to do this, to either get an increase in performance or to maintain it?"

A hard interval workout — like, say, 400-meter repeats — has a large payoff, but it also requires three days of recovery before the athlete has cleared enough fatigue to go all-out again. A runner, particularly one on the wrong side of 30 like Hilary, can really only do two to four hard workouts in a week without building up a dangerous level of fatigue, but that's not nearly enough training to sustain the aerobic base she needs. Hence the polarization solution: complementing those hard workouts with a large volume of work performed at such a low intensity, it can be done even in a fatigued state and creates no additional fatigue of its own. That means going really, really easy, something elite runners almost by definition struggle with. When she trained with a group of Kenyan runners once, Hilary recalls, they ran so slowly in their low-intensity workouts she had trouble staying with the pack and not surging ahead. (No wonder. During recovery runs at altitude on days after a hard workout, Meb would start at a 7:30-mile pace. That doesn't sound so slow, but it's 50 percent slower than his race pace. If I were to polarize my workouts to the same degree, 80 percent of my runs would be walking briskly.) You might wonder how someone who runs a 5K in 15 minutes gets anything out of workouts at two-thirds that speed. But perhaps that's because, as Laird Hamilton says, we're conditioned to believe working out equals feeling beat up.

Whether or not they call it polarization, this principle highlights perhaps the single biggest difference between how elite athletes exercise and how the rest of us do it. They know exactly what they want to get out of their workouts going in, and every workout is part of a bigger, coherent strategy, not a one-off. If you run or bike or swim, unless you're working with a coach or cribbing from professional training programs, chances are you're doing something akin to threshold training: going what feels like pretty hard for what feels like pretty long. Or maybe you're on the HIIT bandwagon. If so, the research suggests you're probably training both too hard and not hard enough — going faster than you should in your casual workouts, and then, when you decide to go for it, going slower than you could because you're not fully recovered. That's probably true even if you think you're mixing in "easy days" or "recovery runs"; unless those easy days are so easy they feel a little weird, they're not easy enough. If you're somebody who's only going to work out a couple of times a week, then sure, max intensity is probably a good way to go. But if you're looking to add more volume to your program but find yourself stymied by injuries or tiredness, or if your progress has plateaued and you don't know why, polarization could be your answer.

Two weeks after meeting the Stellingwerffs, I pull the Subaru into a strip mall in Huntington Beach, one of the low-slung, interchangeable towns between Los Angeles and San Diego. Inside a small, nearly bare storefront office, two middle-aged men invite me to have a seat at a folding table. One of them places Velcro straps around my arms, just above the biceps, and attaches them to a toaster-sized machine, which causes the straps to inflate like blood pressure cuffs. Within a few moments, my hands start to feel tingly, like they're falling asleep, and the flesh of my palms has turned a mottled purplish hue.

"Those pins and needles you're feeling are literally new capillaries being formed," the younger man, a former champion swimmer and coach named Steven Munatones, tells me. "Check your capillary refill." I press a fingertip into the meaty part of the opposite palm and release it, leaving a white circle that quickly fills in with pink as the blood rushes back in. "What we're doing now is a form of warm-up," he says. "We're putting pressure on the vascular walls, then releasing it. Putting pressure on the vascular walls, then releasing it. We're warming you up from the inside out."

I do another capillary-refill check. This time, the white spot lingers. "You're engorged," the older man, an entrepreneur named Richard Herstone, says. "Your arms should start to feel heavy, like you've been doing something." They do, and I haven't. But now it's time to.

Munatones disconnects my armbands from the machine and instructs me to flex my hands, balling them into fists and opening them up, over and over again. I begin. By the time I've done 15 or 20 of these "hand-crunches," my forearms feel as tight as sausages, like they do on the rare occasions I go to a rock-climbing gym.

"We call this 'priming the pump,'" Munatones says. "Lactic acid is building up in there.'"

I'm supposed to do three sets of 25 hand-crunches. Midway through the third set, Munatones looks at me with a clinical eye. "If that's all you can do, that's OK," he says. I feel a little embarrassed. Did I really look like I was having a hard time making a fist? (I was, though.)

We move on to the next exercise: biceps curls. I'm still empty-handed, but Munatones tells me I won't need any weights; I just need to make sure to squeeze at the top of each rep. Again, I'm doing three sets of 25. I work out regularly, but mostly bodyweight stuff, with some smaller weights and resistance bands in the mix. It's been years since I've curled heavy dumbbells. After 15 of these weightless curls, my arms feel fatigued in a way I remember from back then. I notice something else about them: the skin is suddenly speckled with tiny red dots, like freckles. "Those are petechiae," Herstone tells me, little blooms of blood caused by leaking capillaries. "About twelve percent of people get them. Mostly women and fair-skinned people." They go away in about a day, he says. If I were to do this workout repeatedly, I might get petechiae one more time, "and then the vascular walls start to adapt to the strain that we're putting on the body."

Last exercise: triceps extensions, still unweighted. This time, I don't even make it to the second set. My arms are toast. "So, tomorrow, you're gonna feel sore," as after an intense weight-lifting session, Munatones assures me as I loosen the bands.

"But the difference is," Herstone interjects, "you didn't tear any muscle for that feeling."

What I just did is Kaatsu. It's an updated version of a technique developed over the last 50 years by a Japanese exercise scientist named Yoshiaki Sato. Japanese baseball players and American bodybuilders have been practicing Kaatsu — or its generic form, called blood-flow restriction, or occlusion training — for years, but it's only beginning to enter the mainstream of professional sports thanks in part to new hardware that standardizes its application for safety and effectiveness. Along with the AlterG, Kaatsu is one of a number of new technologies that seek to make athletic training more efficient and effective by separating the good kinds of training stimuli — the ones that promote desirable adaptation — from the kinds that cause fatigue, injury, and repetitive strain. Like cross-training, polarized training, and altitude training, these technologies hold particular promise for older athletes, who are more affected than their younger counterparts by those limitations.

As the story goes, one day in 1966 Sato, an amateur bodybuilder and scientist, had just completed a long Buddhist ceremony that involved spending several hours in a kneeling position. In the hours afterward, he noticed that his calves were sore in exactly the same way they were after lifting weights. Guessing that the soreness had something to do with the way kneeling cuts off circulation returning from the limb, Sato began to experiment with various ways of achieving the same effect by wrapping straps and hoses around his arms and legs. Just as he had suspected, performing exercises in a state of partial blood-flow restriction — with the limbs bound tightly enough to compress shallow-lying veins, which return blood to the heart, but not deeper-lying arteries, which carry it outward to the extremities — caused his muscles to balloon. As his investigations progressed from self-experiments to full-blown clinical studies, he began to understand why.

In the classic model of progressive overload, muscles grow bigger and stronger in response to exercise vigorous enough to cause tiny tears in their fibers. That microscopic damage is the major reason you're weak and sore for a couple of days after a hard workout. (It's also why remedies that prevent muscle soreness, like taking large amounts of anti-inflammatories, can diminish adaptation as well.) What Sato suspected, and what his research bore out, is that it's not the microtears themselves that cause new muscle growth; they're merely the usual by-product of another process that triggers it. That process, he determined, was the stretching of capillaries, hair-thin blood vessels that suffuse every living tissue, with walls so thin that oxygen and nutrients can pass right through them. By trapping blood in the limb, Sato was raising the blood pressure there, causing the capillaries to distend more than usual. (The word *kaatsu* means "additional pressure" in Japanese.) By keeping it pooled rather than allowing it to drain back to the heart immediately, he was causing the tissues to marinate in the chemicals released by that vascular flexing, including lactic acid and nitric oxide, which trigger the formation of new muscle. Basically, Sato was tricking the body into thinking it had just endured a bout of hyperintense exercise. Which, if you want to get technical about it, it had. "What exercise actually is is blood flowing to the limb, muscle being contracted and flexed, and the blood vessels expanding and contracting," Munatones says. "The brain doesn't know if I ran, if I pumped weights, if I just sat here. All it knows is it received a signal from the muscle cells that said, 'This muscle exercised.'

"If we can avoid the muscle tearing, but create the same chemical reaction in the brain, we don't have to tear the muscle in order to make it stronger and bigger. So we skipped a step."

Uncoupling muscle building from muscle tearing has some huge benefits. For one thing, it means the recovery profile, that ever-present concern of Trent Stellingwerff's, is almost nonexistent. (My arms were indeed sore the next day, but nothing like they would have been had I lifted to failure with weights.) Moreover, people whose bodies can't tolerate the loading forces of weight-bearing exercise, like post-op surgical patients or the elderly, can still maintain or build muscle. The 32-year-old Tunisian swimmer Oussama Mellouli, a gold medalist at the 2008 and 2012 Olympics, started using Kaatsu after he sustained a shoulder injury in the run-up to Rio. Swimming with the bands on allowed him to dramatically shorten his pool workouts while he rehabilitated. "He's a big man. He needs to maintain his musculature and he can do that with Kaatsu because he's not putting any additional strain on his body," says Munatones.

Josh Saunders, a goalkeeper for New York City FC, used it in much the same way after a complication from ACL surgery in 2013 required a bone graft and ligament reconstruction. Saunders learned about Kaatsu from a sports scientist named Jim Stray-Gundersen. As an adviser to the U.S. Ski & Snowboard Association, he had previously used it to help Bode Miller recover from a herniated disk in time for the world championships. (Stray-Gundersen, who previously worked with the company that makes the AlterG, later moved to Kaatsu Global as its chief medical officer.) Another Kaatsu proponent is John Sullivan, a psychologist and sports science adviser to the New England Patriots and various Premier League teams; Munatones says psychologists and neurologists, who understand the role of the brain in mediating virtually every bodily phenomenon, are quicker to grasp the principle behind occlusion training than strength coaches or physiologists.

That Kaatsu is catching on in America now is largely a product of Munatones's unusual blend of skills. In addition to being an electrical engineer, he's also a champion open-water swimmer who at 52 could still swim the 200-yard butterfly in two minutes, seven seconds, less than nine seconds off the U.S. masters record for his age group. In 2001 Munatones was living in Japan and working for Hitachi while also serving as a volunteer coach for the U.S. swim team. He was the only coach who could speak and write the language when the team came to Tokyo for the world championships that year. After the meets ended, one of the Japanese coaches introduced him to Dr. Sato, a then-53-year-old man whose biceps were as big around as foam rollers. Sato told Munatones about his life's work. Japanese athletes had been using Kaatsu since the 1970s, Munatones learned, and in 1992 the country's Olympic committee had honored Sato, who had personally authored or coauthored over 100 peer-reviewed papers demonstrating Kaatsu's effects, for his contributions to sport. Munatones asked Sato how it was possible something so useful was virtually unknown to non-Japanese athletes. "He goes, 'Oh, two reasons. I don't speak English, and I don't travel outside of Japan,'" Munatones recalls. "I said, 'Good! I do.'"

They agreed to work together. Up to that point, Kaatsu was basically an oral tradition, requiring Sato or another expert practitioner who could tie ligatures at precise pressures. Munatones used his engineering expertise to develop a device that could replicate that effect using pneumatic bands, with software that could remember different protocols and users. In 2014 they formed a company, Kaatsu International, dedicated to marketing the technology outside Japan.

While elite athletes make for great marketing, in its birthplace Kaatsu is regarded primarily as a tool for health care, not sports performance. With a high life expectancy and low immigration, Japan has the oldest population of any nation on earth; more than a quarter of Japanese are over 65. Through something called the 22nd Century Medical and Research Center, the Japanese government supports innovations with the potential to help the country manage its looming demographic crisis, from genomic medicine to domestic robots that perform the functions of human caretakers. Allowing the elderly to live independently and avoid expensive medical interventions is a cornerstone of that effort. "Japan has always been way, way ahead of everybody on longevity, not necessarily because they're smarter, but because the needs of their society dictated that," says Munatones. People who are too frail for other kinds of resistance exercise, he says, can still experience Kaatsu's baseline benefits — "keeping the veins and capillaries nice and elastic, keeping the body strong enough so you're not tripping on wet sidewalks." Doctors in Japan have even begun putting

Kaatsu bands on heart-surgery patients as soon as they emerge
from the OR to encourage severed cardiac blood vessels to regrow,
he says.

Munatones's words would prove weirdly prophetic a few weeks
after our meeting when he suffered a massive heart attack during a training swim. He nearly died in the ambulance on the way
to the hospital. While he lay in a coma, his cardiologist, following
what's called the Arctic Sun protocol, placed him in a state of mild
hypothermia, using cold gel packs to lower his core temperature
to 91.4°F in hopes of arresting any potential brain damage caused
by the hypoxia he experienced while his heart was stopped. Munatones lost a full week of memory to traumatic amnesia, but within
a month of returning to consciousness he was swimming again.
Given that Munatones was already doing Kaatsu almost every day
for years leading up to his heart attack, you might think it would
have shaken his faith in it as a tool for vascular health. In fact, he
says he started doing it again as soon as his doctors gave him clear-
ance and thinks it was vital in helping him avoid muscle wastage
while recuperating. "I believed in Kaatsu before," he says. "Now I
know one hundred percent that it is helpful and healthful, and I
was about as weak as they come after face-planting at my home
only a few months ago."

In the U.S., outside of professional and Olympic sports, Kaatsu
has found several receptive pockets of early adopters. One is the
military, which has purchased more than 60 of the units, which
range in price from $2,000 to $5,000, for use by different branches
of the Joint Special Operations Command. No less than NFL and
NBA teams, the armed forces are continually at pains to find ways
to keep their best performers healthy, and for much the same reason. A paratrooper in the Army's elite 101st Airborne unit represents an investment of about $8 million in training and other
costs, while replacing a SEAL is an expense of somewhere between
$10 million and $15 million for the Navy. Most SEALs have bach-

elor's degrees, and the years of training required means they tend to be old for combat soldiers, around 30 on average, making them prone to exactly the same types of overuse injuries and performance decrements as athletes at that age. The military has long looked to sports science for answers, but it stepped up the effort after September 11 when the Pentagon found itself embroiled in two wars in which special operators on the ground rather than tanks and airplanes or large numbers of lightly trained riflemen bore the brunt of the conflict. Nowadays, much of the academic research into orthopedic injuries, nutrition, and human performance is underwritten by grants from the Department of Defense as it looks to optimize the health of its "tactical athletes." At a big conference like the American College of Sports Medicine's annual meeting, olive drab uniforms are almost as common a sight as team-logo tracksuits.

In 2015 Herstone was in Pearl Harbor demonstrating Kaatsu's use for a group of 25 fitness trainers for the Special Forces. Because commandos spend so much time deployed in the field, they typically prefer exercise programs that can be done anywhere, including absurd volumes of calisthenics. Herstone was getting a strong vibe of skepticism from several jacked-up members of his audience, so he singled out the fittest-looking and most skeptical trainer and asked about his fitness regimen. The man replied: a 10-mile run and 1-mile swim, followed by 150 pull-ups and between 1,000 and 2,000 push-ups, depending on the day. Herstone invited the man to put the bands on and demonstrate his push-up prowess. "He got to thirty-six," he recalls. "He couldn't get off the ground, and these twenty-four other guys are giving him crap, as you imagine they would, and I could see this look of pissed. He won't look at me. So he takes off the bands, walks away, and then he comes back and says, 'You realize you just saved me an hour to an hour and a half a day?'"

NASA has also purchased several Kaatsu units in hopes of finding a better way to keep astronauts healthy for long stretches in space in time for the first manned mission to Mars sometime in the next decade. In the absence of gravity, humans quickly lose muscle mass and bone density, and their circulatory systems become less efficient. Sound familiar? "Zero-G living mimics closely the effects of old age," reads a NASA primer on the health challenges of living in space. "Like astronauts, the elderly fight gravity less. They're more sedentary, which triggers the loop of muscle atrophy, bone atrophy, and lower blood volume." To slow the wasting, astronauts run on a device that's the functional opposite of an AlterG, a treadmill with bungee cords that serve the role of gravity substitute. (In fact, it was for this purpose that a NASA engineer and biomechanics expert named Robert Whalen first developed the idea of a gravity-altering treadmill; it was only later that Whalen's son Sean realized it could be used for deloading as well.) But it's an imperfect solution, and there's reason to believe something like Kaatsu could be part of a more complete one. In particular, NASA is hoping blood-flow restriction might ameliorate the worsening of vision astronauts experience as greater intercranial pressure deforms their eyeballs. Trapping blood in the limbs reduces fluid pressure in the head and trunk, theoretically allowing the eyes time to recover; NASA is testing the effects now.

If that's indeed what happens, it will be one more entry in a long list of what you might call Kaatsu's off-label indications. High-intensity exercise stimulates the release of human growth hormone, and so does Kaatsu; NFL teams that use it report the broken fingers and other small injuries players accumulate throughout the season heal faster in players who use the bands. For the same reason, it accelerates wound healing in diabetics and other people with sores that won't close. Trapping blood in the limbs puts the brain in a state of mild hypoxia, triggering the release of EPO and production of new red blood cells in much the same way as altitude training does. You might think athletes who use Kaatsu to reduce the load on their joints would lose out on one of the beneficial side effects of load-bearing exercise, the increase in bone density, but research published in Japan shows Kaatsu actually increases that. (Since several of Kaatsu's effects have been demonstrated through trials in Japan that haven't been reproduced yet in the U.S., the company has to limit the marketing claims it can make here.)

Hearing such a litany of claims for a health product automatically puts me in mind of a traveling huckster selling old-timey patent medicine: *It relieves constipation, improves cogitation, and restores manly vitality to the nether regions!* The people behind Kaatsu know that. "We only make one claim," says Herstone. "Builds muscle." When you think about it, that's the only claim they need. After all, there's already a miracle drug that increases life span, boosts immune function, releases good hormones and suppresses bad ones, improves sleep, and speeds up metabolism. It's called exercise, and it, too, pretty much just builds muscle. It's just that, physiologically, exercise doesn't have to involve, you know, exercise.

In 2006 Daniel Chao and Brett Wingeier became interested in the question of what happens when you shock different parts of the brain with electrical current. Chao, a physician with a master's degree in neuroscience, and Wingeier, a biomedical engineer, had been colleagues together at a medical device company that developed the first closed-loop neurostimulator, a tiny unit that could be implanted in the brains of people who suffered from epileptic seizures. The device, called the NeuroPace, worked almost exactly like a cardiac pacemaker: when it detected the pattern of bioelectrical activity associated with an incipient seizure, it delivered a small charge of electricity, which normalized the brain activity and averted the seizure. The device won a rare unanimous approval from the FDA, and long-term follow-ups with NeuroPace recipients showed they were having fewer seizures five years postimplantation than one year after. That indicated the device was not only correcting the brain, it was actually teaching the brain to correct itself, unlike seizure medications, which lose their effectiveness over time as the body builds up tolerance.

The experience left Chao and Wingeier with the belief that the biotech industry had only scratched the surface of neurostimulation's power. As effective as the NeuroPace was, most epileptics opted for drugs anyway, reluctant to undergo brain surgery. For a product to have mass potential, Chao and Wingeier agreed, it had to be something that delivered current from outside the body, passing through the hair, scalp, and skull. They waded into the scientific literature, pulling together every study they could find on the effects of transcranial direct-current stimulation (TDCS). They knew the Pentagon's Defense Advanced Research Projects Agency, or DARPA, had been experimenting with TDCS for years, using it to help snipers and pilots get better at picking out camouflaged targets from background clutter. They read studies showing that healthy people and stroke patients were able to develop fine motor skills faster after receiving TDCS. ("The older the subjects, the more prominent this improvement appeared," noted the authors of one such paper, published in the journal *Neurobiology of Aging*.)

Chao and Wingeier built a prototype headset that could deliver current to different areas of the brain, the parts responsible for sensory inputs, language processing, and memory. They reproduced the studies they'd read to make sure they were doing it right, then started conducting new investigations of their own, using "sham" current — strong enough to prickle the skin but too weak to influence the brain — for the control group, to rule out placebo effects. Overwhelmingly, the strongest data they saw came from the motor cortex, which controls the skeletal muscles. After 10 minutes of TDCS to that region, they found, human subjects were able to generate more muscular force and acquire new physical skills faster — and it wasn't even really close. They had what they were after. The partners quit their jobs and, on January 1, 2014, started Halo Neuroscience, based in San Francisco.

At first they were entirely focused on building a device that did what it was supposed to. Chao says they never even thought about who their customers might be for the first year of development. Although he's an avid cyclist, neither of them had any background in elite sports. "I could never have predicted when we founded this company that we would be working with pro athletes, not even in my wildest dreams," he says. Yet even before its first production units went on presale to consumers in March 2016, the Halo Sport was already being used by teams in Major League Baseball and the NBA; by that summer, Olympic track-and-field athletes from five countries would use it in their training for Rio.

TDCS works through an effect Chao calls "neuropriming": the mild current lowers the activation threshold of neurons so that they fire more readily in response to a stimulus. "There's this old adage in neuroscience: neurons that fire together wire together," Chao says. "Neuropriming helps your brain be more plastic, and neuroplasticity is the process by which we build new circuits. Building new circuits is everything for us. It's how we learn everything from new languages to playing the guitar."

If Kaatsu supercharges training by, in effect, changing the biochemical environment of the muscles to trick the brain, TDCS works in the opposite direction. Since the motor neurons are more excited, they send a stronger impulse to the muscles, allowing them to contract more forcefully than they otherwise would be able to. And since they're in a state of heightened plasticity, whatever the muscles do — whether that's producing more power or rehearsing a new skill — they're better able to reproduce later, in a nonprimed state. The brain literally has a muscle memory of being able to do something it couldn't do before. "Let's say you see me at Golden Gate Park and I'm struggling to do a pull-up. You give me ten pounds of assistance and I can start doing a pull-up," Chao says. "We're the ten pounds of assistance."

One critical proof of Halo's efficacy came from a four-week study conducted with seven members of the U.S. Ski & Snowboarding team. The athletes, ski jumpers, trained specifically for the moment of liftoff from the ramp, an event that requires them to produce an enormous amount of force in a smooth and symmetrical but sudden movement. "They have to explode off this surface in a way that's maximal, but also it needs to be a quality movement because it's zero friction," says Chao; a lopsided or herky-jerky jump can easily end in a crash. The skiers performed simulated jumps on a $25,000 force-sensing plate that generates 1,000 measurements per second. Over the course of the four weeks, all seven skiers saw improvements in the amount of force they generated and decreases in the amount of entropy, or deviation from the desired pattern. But the skiers using Halo achieved 13 percent more propulsion and 11 percent more smoothness than the ones in the control group, who used sham devices.

The rapid reduction in neural "noise," or randomness of signal, that happens with neuroprimed training is especially helpful in what's known as stereotyping of a movement, rehearsing and refining it so that it can be performed the same way, every time. When Stephen Curry hits three-pointers from all over the floor, it's because he has stereotyped his angle of release to a precise 46 degrees no matter the circumstances, whether he's alone in space or getting fouled to the floor. Mikel Thomas, a hurdler from Trinidad and Tobago, used Halo in the months before Rio to perfect the timing of his strides and jumps. "The hurdles is an event where you have ten opportunities to fall on your face," he says. "If you can eliminate the amount of mistakes you produce by replicating what you just did, it will produce a successful race."

As with Kaatsu, it's tempting to dismiss claims like these out of hand as too good to be true, the sports science equivalent of those MAKE $5,000/DAY WORKING FROM HOME emails in your spam folder. "You know, almost any intervention we have makes the body work worse," says Marcus Elliott, a Harvard-trained physician who runs a high-tech performance clinic called P3 in Santa Barbara. (We'll get to know him better shortly.) "This idea that we apply this electrical current to the brain and now you are going to be able to learn faster and you are going to create more power, it feels kind of far-fetched and not usually how it works. But there's pretty good literature saying it might work that way." As scientists unlock the mysteries of human biology, maybe we'll just have to update our idea of what's believable.

5

FERRARIS AND TOYOTAS

The Science of Efficient,
Injury-Free Movement

When Peyton Manning wanted to put everything he had left into one last bid for a second Super Bowl title, he put out the SOS to Mackie Shilstone.

Mannning's team, the Denver Broncos, had come tantalizingly close the previous two seasons, losing a Super Bowl to the Seattle Seahawks in 2014 and getting knocked out in the playoffs in 2015 by his old club, the Indianapolis Colts. At 39, Manning knew making it back to the Super Bowl again would make him the oldest quarterback to play in an NFL championship game. Unfortunately, he was starting to play like it. The NFL had its share of ageless wonders, including Manning's archrival, Patriots quarterback Tom Brady. Manning was never one of those. Four neck surgeries to relieve pinched nerves in his cervical spine had left him with one of the weaker throwing arms among starting signal-callers. On top of that, he'd finished the previous season with a torn quadriceps muscle, an injury that had hobbled him and made him an easy target for pass-rushers in the playoffs.

But Manning also knew another Super Bowl victory would change the way he'd be remembered after retirement. Of the players with better overall passing statistics than his, only one, 49ers quarterback Steve Young, had multiple championships to his name. Manning was already a lock for first-ballot entry into the NFL Hall of Fame, but as a two-time Super Bowl winner, he'd at least be in the discussion about who was the greatest quarterback ever to play the game. The good news was that overwhelming physicality had never been Manning's game. Despite his ideal size — six foot five, 230 pounds — he had never been a particularly powerful thrower nor an agile scrambler. His strength had always lain in reading defenses, adjusting play calls to capitalize on their vulnerabilities, and getting the ball out of his hand quickly. Hoping to squeeze a trophy out of his aging superstar, Broncos general manager John Elway, the reigning oldest quarterback ever to win a Super Bowl, had assembled the league's best defense. Elway didn't need Manning to be the second coming of Michael Vick to have a good chance at winning it all. He just needed him to be able to throw the ball well enough downfield and to the sidelines that opposing defenses would have to defend that territory honestly and not cheat up toward the line of scrimmage.

But six weeks into his off-season, Manning had doubts about whether he would be able to manage even that much. So he sent a text message to Shilstone: "I want to know if I can come see you." It was not entirely out of the blue. A longtime New Orleans resident, Shilstone was friendly with Manning's parents, Archie and Olivia, who lived a few blocks away from him in the city's Garden District. After watching Peyton look slow and discombobulated in the blowout Super Bowl loss to the Seahawks, Shilstone had sent a text to Archie, the former Saints quarterback, offering to do whatever he could to help Peyton as a personal favor. "I just don't want to see him end like this," he wrote.

A year and another postseason disappointment later, Peyton

was ready to take him up on the offer. Shilstone received the quarterback's message on a baseball diamond in Arizona where he was running a conditioning program for Major League Baseball umpires. He called back Manning, who asked if he could come and train with Shilstone for a month in New Orleans. "I said, 'Okay, but I'll need to know everything,'" Shilstone recalls.

Coming from almost anyone else, Shilstone's attitude might have come off as more than a little presumptuous. But the Mannings knew his curriculum vitae. In more than 40 years as a trainer, coach, and hard-to-classify all-around performance guru, Shilstone has worked with thousands of professional and college athletes from football, baseball, hockey, tennis, boxing, and other sports. Along the way, he has developed a reputation for doing one thing better than anyone else: taking athletes on the verge of aging out of their sports and helping them tack on a few more years to their careers. And not just extra years, but great ones. When Ozzie Smith, the Hall of Fame shortstop for the St. Louis Cardinals, first visited Shilstone in 1986, he was 31 and complaining of late-season "bat fatigue." His hope was to eke out another three years. After Shilstone put 15 pounds of muscle on Smith, he led all hitters at his position and went on to play another 10 years. He took an opposite approach with Lions left tackle Lomas Brown, whittling him down to 282 pounds to relieve pressure on his knees and allow him to stretch his career to 18 seasons. In June 2016 Shilstone traveled to London for one-on-one work with Serena Williams, who was coming off the third-straight Grand Slam tournament in which she'd inexplicably fallen to a lower-seeded player. For years, Shilstone had worked off and on with Williams, who early in her career lacked the consistent fitness to match her unequaled talent. To ramp up her cardio capacity, he had put her in a swimming pool and made her hold a jug of water above her head while she treaded water. After losing in the Australian Open in January, Williams had canceled plans to meet up with Shilstone in Palm Springs in

March, and her skid continued through the French Open. But after her week with Shilstone in London, Williams regained her form and won Wimbledon, giving her a record-tying 22 titles, and becoming the oldest woman, at 34, to win a major tournament. She would go on to add number 23 at the Australian Open the following January, revealing only afterward that she played the tournament while pregnant.

"They call me the career extender," Shilstone says. Modesty is not a particular trait of his. He's not shy about taking a share of credit for the success of athletes he has trained, and clearly believes he knows more about topics like physiology and nutrition than specialists in those fields. "I'm ahead of the game. I don't have any myopic blinders," he says. At 65, he claims he's in better shape than most of his clients. "I run with all my pros. They're sometimes scared to run with me because I'm in phenomenal shape and I have this incredible recovery," he says. "Ozzie Smith said, 'That Mackie, he may be training five people but he's working out with every one of 'em, and he'll beat me.'"

In person, Shilstone doesn't look like someone who'd command obedience from the likes of Manning and Williams. A smallish, white-haired man with a surprisingly flutey voice, in a blue T-shirt and khaki shorts, he looks less like a fitness professional than a prematurely aged 12-year-old. A flat stomach and stick-straight posture are the only outward clues to what he does. Over iced teas, under a lowering Louisiana sky that would soon open up and overflow storm drains, Shilstone walked me through how he got Manning where he needed to be.

Physiologically, the job of NFL quarterback isn't an especially demanding one, at least not for a pure pocket passer like Manning. (For a scrambler like Russell Wilson or an option back like Cam Newton, it's another matter.) A high VO_2 max or single-rep squat is less important than squishier attributes like core strength and balance. Brett Favre, who came within a play of becoming the

oldest quarterback to reach the Super Bowl in 2010, told me the
formula that worked for him in the last five years of his career was
dialing it back in the weight room, cycling instead of running, and
slimming down. "I'm not an expert and I didn't hire an expert, but
I felt like lifting a lot of weights was just wearing me down," he
says. "I was just tired of the pounding."

But Manning couldn't afford to take it easier. Whereas Favre
had seemingly been made of iron, never missing a game for injury
until the last of his 20 seasons, Manning was more like glass. Shilstone read all his medical files, ordered scans of his own, had three
surgeons examine him. "I said, 'Look, you know one hit, the right
hit, could take you out. So I'm gonna give you a plan.'" Shilstone
put Manning on an exercise bike and had him do sprint intervals
wearing a heart monitor. On a practice field, he put Manning in a
harness and attached bungee cords to it, then had him simulate
taking five- and seven-step drops back with the football, using the
cords first to slow him down and then accelerate him. He assigned
Manning an adapted version of a shoulder-injury-prevention regimen he'd created for Serena Williams, who, after matches, uses
rubber cords for resistance as she performs her various strokes
backwards.

After watching Manning throw, he diagnosed the quarterback's
biggest problem: he was standing up too straight while releasing
the ball. "If you throw tall, you've lost the power in your legs," Shilstone says. Rising from his chair, he demonstrates what he means
on the sidewalk outside Starbucks, miming a quarterback locking
his knees as the ball leaves his hand. "You're throwing all with your
shoulder. So when you get ready to throw" — here he drops into an
athletic crouch — "you gotta sit down, bend your legs." Manning's
height had always been an advantage, allowing him to see over
hulking linemen, but football has changed since he started playing
it. The proliferation of the spread offense has opened up the field,
making it easier for shorter quarterbacks like Wilson and Drew

Brees to succeed. At the same time, defenses have gone smaller and faster, putting a premium on quarterbacks' ability to drive the ball into ever-smaller windows. Manning needed to remake his throwing motion for the modern game — and his 39-year-old arm.

As it happened, Shilstone had just the remedy: a four-foot-long, four-by-four-inch piece of wood he'd bought at Home Depot and turned into a balance beam. In a gym at Isidore Newman School, where Manning had posted a 34-5 record in his three years as starting varsity quarterback in the early 1990s, Shilstone had Manning stand on the beam and attempt to throw along its axis without falling off. Positioned at the far end of the gym to receive his throws, Shilstone watched as, again and again, Manning released the ball straight-legged and immediately tottered off the beam. When he did manage to stay on it, the throws were weak and wobbly. "You're your own worst enemy," Shilstone called down the gym toward him. Gradually, Manning learned to maintain his balance by keeping his legs bent throughout the motion. Soon, Manning wasn't just staying on the beam; he was doing it while throwing harder than he had been while standing on the ground. Instead of generating power from his damaged shoulder, he was deriving it from his legs and hips, using his torso as mainspring, the arm motion a mere follow-through. Feeling the slightly painful smack of a spiral hitting his receiver's gloves, Shilstone called across the gym, "What do you feel now?" He had to ask a couple of times before Manning admitted, "It's effortless." At the end of the month, Manning left for Denver and training camp; the balance beam went with him.

It wouldn't remain quite effortless. Statistically, the 2015 season would end up being the worst of Manning's career. He quickly acquired some new nagging injuries to go with his old ones and reverted more than a few times to his habit of standing up too tall in the pocket and floating passes, too many of which went for interceptions. Shilstone wasn't surprised. It's seldom that easy for the athletes who come see him; if it were, they wouldn't need him. "I get 'em at crossroads in their lives," he says. Just when Manning finally seemed to be settling into a groove, after a dominant Week 8 performance against Green Bay in which he looked like his old self, he tore the plantar fascia in his right foot and limped through the next three games looking worse than ever. He ended up missing five weeks, returning just in time for the playoffs, where he would beat his old nemesis, Tom Brady, and go on to hoist his second Lombardi Trophy. Through it all, Shilstone was watching. Whenever he saw Manning falling back on his bad habits, locking his legs in the pocket instead of sitting down into his throws, he would fire off a two-word text message to the quarterback: "Balance beam."

Picture two athletes. One's movements represent a Platonic ideal of physical grace: fluid, unhurried, economical — effortless. The other's are the opposite: abrupt, laborious, awkward-seeming. Is one of them likelier to have a long career? Actually, there's no need to picture any of this. Just watch some footage of Roger Federer and Rafael Nadal.

With 17 major tournament titles to his credit, Federer is widely considered the greatest male tennis player ever, but among tennis aficionados, that judgment is as much about aesthetics as statistics. Federer plays beautiful tennis. Every stroke is an exemplar of its type, like a diagram from a tennis textbook come to life. He hits with enough pace to hang with the most muscled-up power players, but never harder than he has to; he has a feathery touch on volleys and drop shots that harks back to an earlier era of the sport. His footwork is as good as it comes, getting him just where he needs to be just when he needs to be there. Other players whose soles slap and squeak on hard courts marvel at how quiet Federer's feet are, his steps as light as a dancer's. He moves equally well to either side, a fact Federer has attributed to a youth spent playing more soccer than tennis. "Most people have weaknesses.

Federer has none," as Andre Agassi memorably summed it up. But the most enduring encomium to Federer's movement skills is from a magazine profile written in 2006 by the late writer David Foster Wallace, who was a nationally ranked junior tennis player himself. In the article, "Roger Federer as Religious Experience," Foster Wallace wrote:

> Roger Federer is one of those rare, preternatural athletes who appear to be exempt, at least in part, from certain physical laws . . . The approaching ball hangs, for him, a split-second longer than it ought to. His movements are lithe rather than athletic. Like Ali, Jordan, Maradona, and Gretzky, he seems both less and more substantial than the men he faces. Particularly in the all-white that Wimbledon enjoys getting away with still requiring, he looks like what he may well (I think) be: a creature whose body is both flesh and, somehow, light.

The other quality of Federer's that only tennis savants truly know to appreciate is his durability. When he withdrew from the French Open with a knee injury in June 2016, it was the first time he missed a Grand Slam tournament in almost 17 years; the surgery to repair that torn meniscus, sustained not on the court but while running a bath for his twin daughters, was the first of his career. That's almost unheard of. Despite having played several hundred more matches in his career than any other player on the ATP tour, Federer is the only top player who has never retired in the middle of a match with an injury. From 2003 through 2016, he was never ranked outside the world's top eight men, and usually in the top four.

In any era, this would qualify as a feat of ruggedness on par with the consecutive-games-played streaks of Favre in football and Cal Ripken Jr. in baseball. But it's especially notable that the hardiest player in tennis history plays in a period when his sport is far harder on the body in practically every way than it used to be. Through the late 1970s all of the major tournaments were played either on grass or on clay. Those are what are known in the language of biomechanics as "compliant" surfaces, meaning they transmit less loading force to a player's feet and legs than, say, asphalt, concrete, or hardwood. A compliant surface "is associated with a lower incidence of the overload types of injury," according to a whitepaper by the U.S. Tennis Association. Yet by the late 1980s, asphalt had become the predominant surface on the tour. In the grass-and-clay era, most professionals, wielding wooden- or metal-framed racquets with small sweet spots, played a serve-and-volley style that kept points short. That changed with the rapid spread of carbon-fiber racquets and synthetic strings in the 1980s. The combination of the two, especially the ball-gripping strings that became popular right around the time Federer was coming onto the scene in the late 1990s, allowed players to put considerably more topspin on the ball than they ever had before. Topspin causes the ball to dive after going over the net, so you can put more power into a shot without sending it long or angle it sharply cross-court without landing it wide. Hitting cannon shots from the baseline replaced serve-and-volley as the strategy of choice, and long rallies of 15 or 20 shots went from rare to routine. Tennis, especially men's tennis in the best-of-five-sets major tournaments, became a contest of endurance and court coverage where players competed on fitness as much as skill.

A tennis season is a war of attrition. "Almost every tennis player has some degree of chronic injury somewhere that they're managing through training, through massage, through anti-inflammatories," says John Yandell, a coach and researcher who uses slow-motion-video analysis to understand players' mechanics. The long baseline rallies mean players do far more lateral running than they used to. What's less obvious, Yandell says, is how much more vertical movement the game involves. That's because those spin-loaded shots bounce higher. "Almost every shot on the forehand side in pro tennis is hit with one or both feet in the air," he says. "These guys are having to go up into the air to keep the ball just at chest or shoulder level because of the amount of torque on the ball. They're loading and exploding into the air and then landing on hard courts," hundreds of times per match. While tennis elbow, an inflammation of the forearm tendon caused by repetitive strain, is as common as its name suggests, the majority of injuries in tennis are in fact to the legs and ankles. There's even a lesser-known syndrome called tennis leg, a calf strain that results from landing on one's foot in the follow-through to a serve. For similar reasons, herniated disks are also epidemic among players; not even Federer has been immune to that one.

No one could illustrate the demands modern tennis makes on the body more perfectly than Rafael Nadal. With his pumped-up arms and thighs, he's a visible symbol of the power-hitting era. His game is built on a suffocating defense; no one else covers more ground or chases down more shots that look unreachably wide. More than any other player, Nadal makes full use of the capabilities of those ball-gripping high-tech strings. His weird forehand stroke — it looks like he's cracking a bullwhip or swinging a lasso — imparts so much angular momentum to the ball, it comes off his racquet rotating at 3,200 revolutions per minute. That's about 40 percent faster rotation than the average pro creates, and 20 percent faster than Federer. With so much spin, Nadal forehands explode upward off the bounce, shooting up past shoulder height, out of most players' preferred strike zones; that's the major reason Federer has lost two-thirds of their head-to-head matches.

But Nadal has had nothing like Federer's success avoiding the injury bug. Almost from the start of his career, he has struggled to stay healthy, missing long stretches of time with maladies of the feet, knees, back, wrists, and shoulder. For Nadal, 13 has been the unluckiest number: that's how many Grand Slam appearances he's been able to string together without an injury absence, versus Federer's 65.

That there's some kind of through line from Nadal's all-out intensity and unusual stroke to his physical fragility is plain to anyone who understands the physics of tennis. All the way back in 2005, when Nadal was not yet 20, Agassi predicted Nadal's career would be a short one if he continued to throw himself around the court so violently in every match. "He's writing checks that his body can't cash," he said (a bit petulantly, since Nadal had just beaten him in a match). Yandell points out that Nadal's spin-heavy forehand is the kind of weapon that packs a nasty recoil for its wielder. Imparting that much rotation to the ball with a stiff composite racquet sends a tremendous amount of torque and vibration into the small muscles of the hand and forearm, which have to absorb the force of the incoming shot and redirect it at a right angle vector. "Where do you think the energy from that shot goes?" he says. Meanwhile, the energy used to create all that spin is energy that's not propelling it forward, so for Nadal to hit with the same pace as other top players, he has to swing that much harder.

But what's obvious and what's provable are two different things. Duane Knudson, a professor of biomechanics at Texas State University, agrees it's logical to assume what he calls Nadal's "kamikaze game" carries a very different injury-risk profile than Federer's flowing, light-footed elegance. But that sort of observation is on the level of what Knudson considers "craft knowledge," not science. Of the Federer-Nadal dichotomy, he says, "No one's taken films of matches and digitized their body motion and documented the speed, and even if they did, it would be a retrospective comparison and you couldn't infer any cause or effect from that. It would be a low level of evidence."

Knudson should know. As a member of the USTA Sport Science Committee, he helped write the whitepaper quoted earlier, a coaching manual on tennis technique and injury prevention. That manual contains factoids such as "If the knees are not bent more than 10 degrees in the serve cocking phase, it places a 23% greater load on the shoulder and 27% greater load on the elbow to achieve the same serve velocity." It's tempting to read that as saying a tennis player who doesn't bend her knees enough will eventually injure her shoulder and elbow, but in fact that's an inference too far for Knudson's scientific comfort. The evidence isn't there to support it. The number of instances where researchers have been able to definitively link certain movement patterns to injury outcomes is dispiritingly small, Knudson says. At best, they've been able to show that certain kinds of training programs reduce the frequencies of specific injuries — for instance, hamstring strengthening to prevent ACL tears.

Knudson is an automatic skeptic of anyone who claims to understand the connection between movement and injury at the level of the individual athlete. There's just not enough peer-reviewed academic research in existence to support most of the claims being made, he says. In the U.S., the National Institutes of Health, which supports most basic medical research, is unlikely to approve grants for sports science. To get government funding for biomechanics studies, researchers like Knudson typically have to show relevance to a movement disorder like Parkinson's disease or seek a grant from the Department of Defense. "What's the public-health interest in tennis elbow?" Knudson sighs. "It's tennis elbow, man."

It's hard to argue with that. But just because the academic world isn't making much progress teasing out the connection between movement, injury, and performance in elite athletes doesn't mean no one else is.

Tony Ambler-Wright has a photo he likes to show students of his graduate course in exercise science. It's an image of the quarterback Robert Griffin III, taken in February 2012 at the NFL

Combine, the annual event where scouts watch top college football prospects perform various feats of athleticism a few weeks before the draft. Griffin, the previous season's Heisman Trophy winner, is in the action of performing what will be a 39-inch vertical leap. That will go down as the highest jump by any quarterback and the eighth-highest for any of the 328 players invited to the combine that year. To the assembled coaches and scouts, it will confirm the predraft buzz on Griffin: a truly special athlete, one whose explosiveness makes him a threat to score on any play. That reputation, combined with a full set of more conventional quarterbacking skills, will make Griffin the second-overall pick in the draft, just behind Andrew Luck. (Seventy-three picks later, the Seattle Seahawks will take another quarterback named Russell Wilson, who, despite similar athletic traits, jumped five inches less than Griffin and required an extra one-fifth of a second to finish the 40-yard dash. That tells you what a rare athlete Griffin was.)

In the photo, however, Griffin isn't airborne yet but squatting, loading his legs for the jump, and it's a strange-looking squat because his knees are practically touching each other. This knock-kneed position, called valgus, is a known risk factor for ACL tears, one of the few reliable pieces of knowledge about a notoriously capricious injury. "Understand that the knee is primarily a hinge joint," Ambler-Wright says. "There's a little bit of rotation available, but this is not a very good position for the knee ultimately." Indeed, Griffin tore the ligaments in his right knee during his sophomore year and went on to obliterate them again during a playoff game 11 months after the combine. But to Ambler-Wright, the photo is not only the red flag all those NFL scouts somehow missed. It's the Rosetta stone for explaining the connection between movement, injury, and career longevity.

I'm looking at this photo on a computer screen in the Atlanta headquarters of a sports medicine and performance company called Fusionetics, where Ambler-Wright works. Fusionetics was started in 2012 by a physical therapist named Micheal Clark, who was head of sports medicine for the NBA's Phoenix Suns from 2000 to 2015. Remember when I said the San Antonio Spurs have been the healthiest team in basketball since the league started keeping accurate injury statistics? There's actually one other team that has lost approximately the same number of player-games to injury over that span. That would be the Suns, and like the Spurs they've done it year after year with teams built around cores of older players. Before capping off his career with the Lakers, two-time MVP Steve Nash led the league in assists and was named an All-Star at age 38 despite a congenital spinal condition called spondylolisthesis, which causes chronic nerve pain in his back and legs. Grant Hill joined the Suns in 2007, when he was 35; after an early career marked by persistent foot and ankle injuries, he proceeded to play 283 of a possible 286 games for Phoenix. Compared with the average team, the Suns have wasted $7 million less per season on paychecks to players who sat out injured, an even better record than the Spurs'.

The Suns credit that success largely to Clark and his methodology for preventive therapy and training, a system he trademarked as Fusionetics, which has since been adopted by half the teams in the NBA and a handful of NFL franchises as well. Kobe Bryant says it was his Fusionetics regimen that allowed him to make it through his agonizing final year in the NBA, when his damaged back, shoulder, and Achilles all threatened to keep him off the court. Amar'e Stoudemire told me it was Clark's program that allowed him to return to a high level after knee-cartilage surgery that seemed to do more harm than good. At the core of the system is the idea of movement efficiency. Every athletic action — Peyton Manning's throw, Rafael Nadal's forehand, Robert Griffin's vertical jump — involves what's called the kinetic chain. That term refers to the way different parts of the body coordinate to produce movement, transferring energy to each other via the joints and connective tissues. A properly functioning kinetic chain is like Karl Marx's ideal society: from each body part according to its ability, to each according to its need. The big muscles in the legs and core provide the serious power; the smaller muscles in the extremities receive that power and translate it into finely controlled actions. When you see an athlete like Federer making 130-mile-per-hour serves look effortless, what you're seeing is a seamless kinetic chain. When the chain breaks, however, that smooth transfer of power gets interrupted and the weaker muscles and connective tissues have to work harder than they should to make up the difference. That's when you get Peyton Manning "arming" the ball and throwing a wobbly duck. Movement efficiency basically just means how smoothly an athlete's kinetic chain operates in performing all the different movements he or she has to do.

When Clark first began working with NBA players, he saw a league full of athletes who were superbly fit but got hurt all the time and didn't move especially well even when they were supposedly healthy. Compared with players from previous eras, they were far more jacked and had higher VO_2 maxes, but that didn't translate into higher-quality movement. When Clark examined them, he found almost every player exhibited less than the normal range of motion in his ankles or hips. He suspected the interplay of the two — extreme fitness and considerable movement limitations — explained why injury rates in the NBA continued to increase even as teams modernized their strength programs and medical care. Adding explosive strength to a body that moves poorly is "like having a Ferrari engine with a Toyota frame and Toyota brakes," Clark told me. Something has to give.

There's a reason Clark found so many players with mobility deficits. Like tennis, basketball is hard on the body, and the life of an NBA player is one of relentless pounding and not nearly enough time to recover from it. A basketball player can temporarily lose as much as 30 percent of the range of motion in his lower-body joints in the day or so after a game. Without adequate recovery or treatment, those reductions pile up throughout a season and from season to season; the longer an athlete has been competing, the more range-of-motion limitations he or she usually displays. In basketball players, according to Ambler-Wright, the ankles are the worst problem zone. Jumping and landing on hardwood dozens of times a night, night after night, produces calf muscles that are chronically tight — "hypertonic," in physio-speak. Tight calf muscles make it hard for the foot to bend upward toward the shin, or dorsiflex. Ankle sprains, the most common injury in basketball, can also create extra stiffness even after they've healed. The vast majority of basketball players have severely limited ankle dorsiflexion. "When we look at the ideal as being twenty degrees, we see elite guys with five degrees or less all the time," he says. "Sometimes less than zero. They're actually negative."

When we think of elite athletes, we usually picture people whose bodies do everything better than everyone else's. In fact, in some key respects, their bodies function worse than the average fit civilian's, such is the wear and tear they sustain all the time. The life of a high-level competitive athlete does not make for blooming health. There's far too much training, competition, and travel without adequate rest, too many injuries that don't have time to heal before reintroduction of the stresses that caused them. On tests of movement efficiency, "the majority of our general-population clients may score better than some of our elite athletes," says Ambler-Wright. That doesn't make the crew from your Tuesday-night pickup game better basketball players than LeBron James, of course. It just means they don't have to overcome the same limitations. "The thing about pro athletes is they're so good at being able to compensate, it doesn't matter," Ambler-Wright says. "That's just how good these guys are."

But compensation is a bargain with the devil. The science of kinesiology comes down to two words: mobility and stability. A joint possesses mobility when it can move in all the ways it's supposed to; it has stability when it moves only when and how it's supposed to. A lack of either mobility or stability is the sort of thing that can cause a break in the kinetic chain. When a basketball player jumps to elevate for a shot or snag a rebound, his knees need to dip toward the floor in order for his thighs and hips to load. If the ankles won't bend, the knees must find another way to get low — for instance, by caving in toward each other, into valgus position. That's a compensation. A little tightness in the ankle is the sort of thing that's easy for an athlete to ignore or fail to notice. But when you know it's a risk factor for an ACL tear or a torn hamstring, that ankle tightness takes on a whole new significance.

To show me what all this looks like in practice, Ambler-Wright puts me through a Fusionetics evaluation. It starts with a modified functional-movement screen, a series of exercises — squat, push-up, plank, lunge — designed to highlight limitations in my strength and flexibility. He instructs me not to concentrate too much on my form. "Just let your body do what it wants to do," he says when he sees me straining to keep my knees over my toes like I know I should. "That will give us insight into what muscles might be too tight or not working hard enough." After I perform the squat slowly, arms extended over my head, Ambler-Wright has me do another one standing with my heels on an inch-thick board, which makes it much easier. ("Basically what we did is provided you with some artificial range of motion in your ankle," he explains.) Then I lie down on a table and Ambler-Wright bends my limbs this way and that, each time measuring the angle of my joints with a protractor-like instrument and entering the results on a digital tablet.

When we're done, he brings my results up on a TV monitor: an outline of a man's body with little red, yellow, or green flags attached to each part. Most of the flags are red. I don't need a degree in kinesiology to know that's not good. A red flag means movement efficiency of less than 50 on a scale where 100 is perfect (or green) and 75 is acceptable (or yellow). My overall movement-efficiency score is a dismaying 41.3. My shoulders, hips, and left ankle are all red, indicating range-of-motion limitations associated with a heightened injury risk. That's not terribly surprising. Even at my fittest, I'm not what you'd call a fluid mover. Physically, I take after my father, whose rare attempts at athleticism call to mind a bundled-up toddler in a too-tight snowsuit. Plus, at the time of the test, I'm rehabbing a passel of injuries, from a strained rotator cuff to lingering weakness in my left leg from herniated disks. I am surprised by the failing score for neck mobility, mostly because I've never thought of neck mobility as a thing. Apparently I turn my head about as well as Michael Keaton in his rubber Batman suit. The deficit is especially bad on the side with my bum rotator cuff; on that side, I scored a perfect zero. Could those be related? "It's all connected, my man," Ambler-Wright says.

Now for the treatment. Ambler-Wright starts with 20 minutes of manual deep-tissue work, pressing his thumbs, knuckles, and elbows aggressively into different spots in my hips, shoulder, and calves and holding the pressure until I feel the muscles underneath stop fighting it and soften. Then he has me do what he calls muscle activations — holding my knees a few inches apart while I attempt to squeeze them together, for instance. "When you have tightness on one side of a joint, you're going to have relative weakness on the other side of that joint," he explains. "We release tight muscles first, and then we activate opposing muscles on other side of the joint in order to reestablish that force balance. If you release a tight muscle and you don't teach the body how to control that new range of motion, it's just going to tighten back up again."

Finally, Ambler-Wright gives me my customized program: 10 minutes of daily stretches and exercises intended to improve the range of motion and stability in my shoulders, ankles, and hips. The exercises are relatively easy ones, the sort of thing you'd do in a physical therapist's office, not a gym, involving little or no resistance. But if I put in 70 minutes a week, Ambler-Wright promises, I won't just reduce my risk for injuries; I'll also get faster and stronger, without running more or lifting heavier. That's because right now, my limitations in those areas make me like a car that's driving with the parking brake still engaged. "The goal of an athlete should be to be as efficient as possible, which means you're expending minimal energy but getting maximal output," he explains.

That's worth thinking about. As they get older, athletes get a little slower and weaker. Some of that is the result of muscular decline, which is pretty much inevitable. But some of it is the result of impairments and compensations accumulated over the long years. The muscles are still as powerful as they need to be; their power is just not making its way down the kinetic chain. Muscle-function decline can be slowed but not stopped, and counteracting it by working harder is tough because older muscles also need more rest between workouts. Mobility impairments, on the other hand, represent low-hanging fruit. Compared with slapping more plates on the squat bar, stretching out your calves and ankles takes a lot less effort, requires no recovery afterward, and can have the same functional effect. For any athlete, but particularly for older ones, the shortest path to getting faster and stronger may be tapping into the speed and strength you already have but aren't able to access. "We're not saying you shouldn't do things to get bigger, stronger, faster, more explosive," says Ambler-Wright, "but if you're training for all these things on a bad frame, on a body that's not moving optimally, you're limiting your performance."

With apologies to Micheal Clark, no one knows more about the science of how basketball players move than Marcus Elliott. That's partly because of how many of them he has studied: starting in 2012, his company, P3, has done assessments of every college player entering the NBA Draft at the league's invitation. It's also a function of how he studies them. At his California clinic Elliott sticks small reflective dots on his subjects and has them perform jumps, lunges, and other maneuvers on $30,000 plates that measure the ground-reaction forces they generate while 12 high-speed infrared cameras capture the precise movements of each body part in three dimensions. Software algorithms then churn through all those data points, allowing Elliott and his team to determine how a player's movement quality compares with that of his peers, where there's room for improvement, and where there's evidence of a hidden or impending injury. "Nobody has collected this level of data on professional athletes," Elliott says. "I mean, nowhere in the world. It just doesn't exist. This was all new when we started."

The motion-capture system at P3 (whose name stands for Peak Performance Project) is similar to the one used to turn actors' performances into computer-generated characters for Hollywood films and computer games. Being able to render an athlete's movements as math and analyze them in the language of physics is crucial because elite players aren't just the best at compensating for their weaknesses; they're also geniuses at disguising them, even to an expert observer. "So often we'll look at slow-motion video of a guy's movements and they'll look symmetric," Elliott says. "And then you look at the numbers and you see a huge asymmetry. Your eyes lie to you. Our sensory systems weren't built to interpret at this level. I mean, we're great at picking up emotions in people, deciding if someone is trustworthy or not, right? We're good at that stuff. But deciding if someone has some asymmetry on how they load on a jump, we're not very good at that."

Elliott and I are having this conversation behind a row of glowing monitors at the back of his clinic, at a console overlooking the force-plate-and-camera setup. P3 is situated on a sunny side street in sleepy Santa Barbara, an hour's drive up the coast from Los Angeles. An avid surfer, Elliott moved his family here for the idyllic lifestyle and trusted the clients would come to him, which they have: not just basketball players but footballers, snowboard-ers, pole-vaulters, you name it. Across the street, there's a shop that does stand-up paddleboard rentals; just down the road is the beach, where you can see parasails dotting the horizon. The clinic itself looks and feels a bit like the spa at a luxury hotel. The young, fit staffers are all clad in stylish black athleisurewear, as is Elliott himself, who, at 48, has the chiseled features and tousled hair of an unrealistically handsome surgeon on a soap opera. In fact, he is a doctor, with an MD from Harvard — the kind of doctor who spends his vacations doing self-imposed personal growth challenges like paddleboarding across the ocean with Cleveland Cavaliers star Kyle Korver.

It would be easy to hate him except that Elliott is, on top of it all, remarkably nice and sincere about his desire to help people. That's a legacy of the experience that led him to medicine: when he was 17, Elliott was playing wide receiver in a high school football game when a low tackle took out all the ligaments in his knee. "I just heard *pop, pop, pop,* like guitar strings, one after the other," he recalls. The injury sent him into a deep depression that lasted most of the next year, but when it lifted, Elliott was transformed from a mediocre student into a man on a mission. "I just came out of it with this intense focus," he says. "I didn't party when I was in college. I was just gonna kick ass, learn everything I could. I was just like, I need to make sure that doesn't happen to more people than it has to happen to."

If Fusionetics is fundamentally in the business of identifying and removing obstacles to efficient movement, P3 focuses on a related but different phenomenon: deeply ingrained movement patterns. "One of the things that slaps us right in the face from the ridiculous amount of data we collect is compensation patterns from old injuries," Elliott explains. "Professional athletes might be great movers and really successful in whatever league they play in, but we'll see something that's just super atypical and pathologic and isn't built to survive. So often it's a new movement pattern that's a result of some fairly recent injury from the last handful of years.

"Almost anybody who's pretty active with their body has had some type of injury in the past that is causing some type of injury now," he continues. "Oftentimes the compensation patterns we develop as civilians don't manifest until your late thirties or your forties. But for these guys, because of the load they put on them, they manifest much more quickly. Everything plays out in a much shorter time period for these professional athletes. Getting these compensation patterns out of their systems, not letting them continue to fester or cause career-limiting or career-ending injuries, is absolutely one of the keys to optimizing athletes deep into their careers."

To illustrate his point, Elliott cues up a video of Detroit Pistons center Andre Drummond, the ninth-overall pick in the 2012 NBA Draft. "This kid should be the best big man in the game for a while if we can keep him healthy for a long period of time," Elliott says. "He's a guy who in a lot of ways moves like a small man though he's almost seven feet tall and two hundred and ninety pounds or so. He's a physical beast." In slow motion, Drummond is performing a vertical-drop jump. He begins by standing on a box 18 inches tall, hops off, and lands in a crouch, then, in a fluid continuation of the same action, launches himself as high in the air as he can. Digitally overlaid on Drummond's body as he performs this action are colored lines showing the various vectors of force acting on each part of his body: compression, torsion, shear. On the other half of a split screen, a graph with two lines shows the ground-reaction force he's sending into the force plates with each leg. Tracking each other closely, the lines peak as he reaches the bottom of his landing crouch and begins exploding toward the ceiling. That peak impact, Elliott says, is what we're looking for. In amplitude and tim-

ing, Drummond's peaks are within 5 percent of each other. Among
NBA players, the standard deviation is 14 percent, which makes
Drummond "super symmetrical."

Then he loads another video. To maintain the subject's confi-
dentiality, Elliott shows me not the raw footage but a version in
which the player is rendered as a skeleton. This athlete, he says,
was one of the best college basketball players coming out, a lottery
draft pick beginning his NBA career. The skeleton performs the
same drop jump, only his force graph looks completely different:
the line showing the ground-reaction forces on his right side spikes
steeply but the one measuring his left side is just a low, rounded
dome. "It's like his body doesn't know he's hitting the ground right
now," Elliott says. "And this is a guy who says he's perfectly healthy
when he comes to us."

A slow-motion replay of the animation shows the reason for
the discrepancy: on impact, the jumper's left heel never actually
touches the ground. Meanwhile, at the low point of the jump, the
right side of his body becomes supercompressed, his ankle buck-
ling deeply and his right knee wobbling briefly outward, then
crossing back over the midline into valgus. "You can see all the
load is going through his right side," Elliott says. "That's not built
to last. There's no way he's going to survive doing this."

After recording this series of jumps, Elliott presented it to the
player himself, who confessed that, contra his claim of perfect
health, he had in fact been experiencing knee pain — enough that
he had secretly obtained an MRI without informing his new team.
The scan showed no significant damage, but when Elliott elicited
a full history, he learned the player had toughed out a bad case
of plantar fasciitis in his left foot during his senior year. The foot
had long since healed completely. "He's got zero pain, there's no
scar tissue, there are no obvious anatomical remnants of this injury," Elliott says. "But, biomechanically, they are all over the place.
The traces are all over him." Elliott created a training program to

reteach the rookie how to jump and correct his various deficits; within five weeks of his assessment, a 40 percent discrepancy in force output between right and left was down to 10 percent, within the NBA normal range.

Looking at movement patterns rather than simply at movement limitations makes sense because the human body is not, in fact, a car. A machine is deterministic: if you take a wheel that's out of alignment and straighten it out, the tread won't wear unevenly anymore. People, on the other hand, have habits, tics, sense memories. If you remove all the mechanical constraints on efficient movement but don't retrain the neuromuscular patterns that developed under those constraints or produced them in the first place, the athlete will be prone to the same issues again and again, Elliott believes.

Moreover, not every pathological movement pattern has an obvious basis in compensation. After analyzing hundreds of basketball players' drop jumps, Elliott and his research team realized that a full one-third of them exhibited a markedly different pattern than the rest. A jump consists of two distinct phases: the eccentric phase, where muscles are lengthening and loading, and the concentric phase, where muscles are contracting to produce explosive power. In this subgroup of jumpers, however, the phases weren't distinct at all. Instead, they overlapped, with the glutes and lower-back muscles contracting to elevate the torso while the quads and calf muscles continued to lengthen into a squat. Elliott shows me a video of one subject as an example. "It's almost like he's trying to jump out of his hips while trying to land out of his knees," he remarks. They nicknamed the members of this group "blenders." Cross-referencing jump data with medical-intake forms, the researchers discovered that the blenders were 300 percent more likely to have back injuries than the other type of jumper (termed "yielders"). Since the data was retrospective rather than the product of a prospective study, Elliott can only say for certain there's a correlation, not a causal relationship. But if he can't conclusively say it causes back injuries, there's also no reason to think it results from them, either. "I don't know if we can even say it's a compensation pattern. It's just how a third of these guys move," he says. "It might be that it served them at some level — they get off the ground faster this way."

Here's where the work at P3 departs from strict evidence-based science into the realm of intuition and expertise — what Duane Knudson would call craft knowledge. There may be no ironclad proof that blended jump mechanics cause back injuries, but Elliott believes the data are compelling enough to act on. The convergence of force vectors in his 3-D models makes it obvious to him that blended jumping can't be good for your spinal disks, so why not train athletes out of it? "There are some clear Newtonian physics behind this stuff that everybody has to accommodate," he says. "There's a reason you've had two diskectomies. That's not a random thing, there's some physics behind that. There's some way that you use your body and it's related to how your system is put together."

If P3 is guilty of treating athletes with a methodology based on partial evidence and theory, so is the traditional sports medicine establishment. When a nonathlete sustains a herniated lumbar disk, Elliott points out, 9 times out of 10, her doctor or physical therapist will prescribe core-strengthening exercises. That's not mere voodoo; there's plenty of research connecting back pain to weak abdominal muscles. But it's fundamentally a public health approach, the easiest intervention that will benefit the greatest number of people. At the level of the individual, the public health approach breaks down, especially if the individual is an elite athlete. Their bodies are subject to very different demands, and their health is too precious a commodity to be entrusted to answers that work for most of the people most of the time.

Prior to the 1960s, the conventional wisdom in high-level sports was that weightlifting was something most athletes ought to avoid. It was too easy to get hurt or become "muscle-bound" — slow and inflexible. That prejudice rapidly fell away as teams in the NFL and other sports began strength training with excellent results, and a vast body of research showed well-constructed strength programs helped prevent injuries rather than causing them. But Elliott, like Micheal Clark, has seen enough to know that more strength is almost never what truly elite athletes need. Most of them are plenty fit enough to do their jobs. The issue is the quality of their movement. "It's rare to find a basketball player without a six-pack," he says. In a blender with back pain, he says, "it's not about making his core stronger — it's that every time he lands he creates so much shear in his lumbar spine that it has no chance to survive no matter how many crunches and bridges he does. This stuff is about what are you asking your body to do and how equipped is it to do that. Up until now, we just haven't measured humans in a high-enough volume to get these kinds of insights."

6

THE DESTINY IN YOUR CELLS

What Genes Have to Say
About Finishing Strong

Look at this!" Stuart Kim exclaims. "That's either very good news or very bad news."

Kim, a Stanford University professor who studies the connection between genetic inheritance and aging, is poring over an intriguing batch of DNA data on the computer screen in his Palo Alto study. It's a list of single-nucleotide polymorphisms, or SNPs, known to be associated with healthy aging and longevity in humans. SNPs (pronounced "snips") are one-letter variations in the three-billion-character novel that is the human genetic code. Researchers like Kim have identified more than 10 million of them that occur in at least 1 percent of the population, and are rapidly untangling the mystery of how they influence susceptibility to genetically influenced diseases like cancer and Alzheimer's, or variability in traits like height and caffeine tolerance.

What's grabbed Kim's attention in this case is a curious repetitive pattern in the data before him. Since every person, barring genetic disorders, receives a full set of chromosomes from each parent, a SNP consists of two nucleotides, the organic molecules that make up DNA. There are four nucleotides in DNA, represented by the letters *A, C, G,* and *T,* so most SNPs consist of two different letters. But in the table on Kim's screen, in row after row, the SNPs are "homozygous," or matched pairs: AA, GG, TT, AA, CC, and so on. If an allele, or gene variant, is one that has some beneficial function or confers some protection against disease, having two copies of it can be, as Kim says, very good news — but if it's one associated with an increased risk, then it's doubly bad to be homozygous. Kim is capaciously knowledgeable and obsessed with his area of research — "As a hobby, I screen through all genetic studies for all diseases," he mentions in passing — but even he doesn't carry around in his head a list of all 10 million known human SNPs and their effects. For the moment, then, all he can say is whoever provided this DNA sample stands a considerably better-than-average chance of surviving into old age in good health. Or maybe it's a considerably worse chance. Definitely one of the two, though.

The subject, of course, is me. And now I'm a little nervous.

In preparation for this conversation, five weeks earlier I had drooled a few cubic centimeters of saliva into a plastic vial and mailed it off to a company called 23andMe, based in Mountain View, California, just down the road from Kim's lab in Palo Alto. For 200 bucks, 23andMe analyzed my SNPs and sent me back a report full of fun but mostly useless facts about myself. I learned that I have more DNA in common with now-extinct Neanderthals than two-thirds of *Homo sapiens* do and that I don't have a gene variant that causes some people to sneeze when exposed to bright light. What I didn't learn was much of anything about how I might expect aging to affect me, athletically or otherwise.

As it happens, 23andMe's founder, Anne Wojcicki, is pretty much obsessed with healthy aging and physical fitness. When I met her at her startup's offices, a few months before my conversation with Kim, she was wearing jogging shorts, a tank top under a zip-up sweatshirt, and running sneakers. She had just come not from the gym but a meeting; workout clothes are her daily uniform. "I'm always ready to go for a run," she told me. That run would be in addition to the twice-daily workout she gets riding an ElliptiGO — that bike/elliptical trainer hybrid Meb Keflezighi used for cross-training — the thirty minutes to and from her home in Los Altos. Those rides are her version of meditation; she always arrives at the office with at least one good idea born on the road. At work, she only ever takes the stairs to and from her fourth-floor office and gets annoyed when she sees employees riding the elevator. "We're a health care company. It's only four flights," she says. In addition to her gym togs, Wojcicki sported two fitness-tracking watches, one on each wrist, which she was testing to determine which one recorded her heart rate more accurately.

A few weeks before our meeting, Wojcicki had noticed herself getting winded on the stairs. When she checked her heart-rate data, it indicated she was in the same exemplary shape as ever, so she asked her doctor for a blood test, which revealed she was "super anemic" and in need of iron supplements. That's the kind of "actionable" data-driven insight Wojcicki had in mind when she started 23andMe in 2006. The company quickly made headlines when Wojcicki's then husband, Google cofounder Sergey Brin, announced his test had revealed he had inherited a mutation that gave him a 50 percent chance of developing Parkinson's disease, a chronic and progressive neurological disorder. In interviews, Brin has said he believes his own formidable fitness regimen — he does gymnastics, yoga, and diving — will improve his chances of staying healthy. He has further stacked the deck by donating tens of millions of dollars to Parkinson's researchers and putting some of Google's money toward a longevity-science startup called Calico. (Wojcicki and Brin had just gotten divorced when I interviewed her; a few months later, she began dating New York Yankees superstar Alex Rodriguez.)

But the federal Food and Drug Administration soon stepped in and ordered 23andMe to stop giving users reports like the information that changed Brin's life. In 2013 the agency officially prohibited the company from telling customers about their genetic risk factors for diseases such as Parkinson's and cancer, citing two reasons. The first was accuracy: rather than sequence the entire genome, 23andMe essentially conducts spot-checks of certain SNPs and then fills in the rest with educated guesses based on probability. That method left too much room for error, the regulators said, considering a customer might well respond to a false-positive finding of a breast-cancer-causing BRCA1 gene mutation by having a prophylactic mastectomy. The company was also deemed guilty of short-circuiting the role doctors are meant to play in interpreting complex medical data for consumers who lack the tools to understand it. After lengthy negotiations with the feds, 23andMe was able to resume issuing certain types of health reports, but it has to be very careful about anything involving predictions of future health or disease.

To learn anything worth knowing about my individual genetic destiny, then, I was going to have to go a different route. The question in my mind was: What does everything we know about the science of aging, and the outsize role genetic inheritance plays in that process, tell us about how we age as athletes? Are there clues in our DNA that explain why one sports star peaks at 25 while another remains world-class into her early 40s? Are there hints in mine as to why my body has aged the particular way it has and how much longer I can expect to feel strong and (mostly) pain-free?

It didn't take much reconnaissance to determine that Stuart Kim was the person I needed to speak with. There wasn't a long list to choose from. To say there aren't many geneticists who have done extensive research into the roots of both life span and athletic ability is putting it mildly. "I'm pretty sure I'm the only one," he told me in our first meeting.

In his lab at Stanford University, where he holds a professorship in the Department of Developmental Biology, Kim and his team of grad students use genome sequencing and genome-wide association studies to hunt for the bits of code that control different aging processes. They study how aging manifests in organisms from the minuscule roundworm *C. elegans*, whose simplicity and rapid reproduction make it an ideal subject for lab experiments, to human centenarians and supercentenarians. The latter, people who live past 110, are exceptionally rare; at last count there were only 22 in the United States. "I'm trying to figure out why people who get to be a hundred get to be a hundred, and I work on model organisms to try to figure out what controls life span," Kim told me on a visit to his lab.

That elite athletes live longer, on average, than the rest of us is well established. Just how much longer depends on the kind of athlete. Studies have found professional baseball players enjoy life expectancies four to five years longer than other men born at the same time, football players live an extra six years, and Tour de France cyclists enjoy an eight-year life span boost. Athletes in endurance sports or sports that demand a mix of endurance and power (e.g., soccer or basketball) fare better than pure power athletes, like weightlifters. High-impact collisions like those in hockey, football, and rugby decrease a sport's longevity effect, as the concussion crisis in pro football has unfortunately made clear. Within sports, players with high body-mass indexes, like football linemen, tend to die younger than those with lower BMIs.

None of that is terribly surprising. We know fitness has a large effect on longevity, so it's logical to assume people whose jobs demand high levels of one would experience more of the other. A more interesting question is whether athletes who live longer also enjoy longer physical primes. Again, the logic lines up. Among researchers on aging, there's a theory gaining credence that longevity is a result of slower maturation across an organism's entire life span, not just at the end of it. A key piece of evidence for that hypothesis involves reproduction: women with longer-than-average life spans are more likely to have experienced puberty and given birth later than their peers.

Meanwhile, there's evidence that the later an individual matures, the more likely he or she is to achieve athletic greatness. A 2013 Indiana University study of track-and-field athletes found that elite junior and senior athletes comprised two distinct groups, with those who achieved their physical peaks before age 20 typically not continuing to be top competitors after that age. (The researchers' takeaway was that national sports associations charged with cultivating the best youth talent should be wary of placing too much stock in early performance.) Since pro athletes tend to be late bloomers, could that be connected to their tendency to live longer? And wouldn't that suggest, all other things held equal, they would maintain their athletic characteristics longer than the early-developing and short-lived? Are the rates of decline in athletic characteristics like VO_2 max and muscle contractile speed a function of the overall rate of aging, or are they governed by specific genes?

Those are the kinds of questions that interest Kim. Until more scientists join him at the intersection of sports performance and longevity, however, they're likely to remain just questions. Any relationship between athletic longevity and the other kind, and the existence of mechanisms that connect the two, is strictly hypothetical. "You'd think somebody who ages slowly and lives a long time might also be an athlete for a long time," he says, "but I don't know of any data that would support that."

Kim himself was only looking at one side of the equation until 2008, when he had a fateful lunch with Jim Kovach, a retired linebacker for the New Orleans Saints and San Francisco 49ers. At the time, Kovach was president of the Buck Institute for Research on Aging, which is located in Marin County, just north across the

Golden Gate Bridge from San Francisco. Kim was on the institute's scientific advisory board. The two got to talking about how genetics, Kim's area of expertise, influences athletic traits like the size and speed that caused the Saints to select Kovach with the 93rd pick of the 1979 draft. Kovach offered to round up 100 of his ex-NFL buddies to provide samples for a study if Kim would do the analysis. "The idea was, if you take the biggest, strongest people on the team and you compare them to just normal people, do you see genes that might tell you about why they're genetically big and strong?" Kim says. They roped in another geneticist, Huntington Willard of Duke University, and persuaded 23andMe to donate 100 "spit kits" and perform the genotyping.

It was easy enough to get the football players to provide drool samples. Convincing them to care about the results turned out to be the hard part. While scientists might be endlessly curious about the underlying basis of athleticism, to the athletes themselves, it was useless information. Kim realized this after going over test results with a man who had been a starting guard for the Cincinnati Bengals. Surprisingly, the man's DNA test showed he lacked the allele for making the protein alpha actinin-3, which is present in fast-twitch muscle fibers. World-class sprinters almost always have two copies of the active form of the ACTN3 gene while elite endurance athletes are considerably more likely than everyone else to have two copies of the inactive form. (Most of us have one of each.) NFL linemen operate in three-to-eight-second bursts of extreme power, much more akin to sprinting than marathon running. Kim told the guard his genetic profile suggested he should have been more suited for a different sport. "He just didn't care," he says. "He knew how strong he was, he knew how many bench presses he could do, and it didn't matter what the genetics were."

But there was something Kim could tell the athletes that they didn't already know. In addition to looking for markers of athleticism, he had also cross-referenced their genotype reports with

SNPs that had been flagged as correlating to a higher- or lower-than-average risk of various common sports injuries. When he got to that part of his explanation, Kim found, almost every player sat up straight and started asking questions. No one was more intrigued than Hoby Brenner, a tight end for the Saints from 1981 until 1993. Brenner, Kim found, had a rare form of the gene that codes for the protein COL1A1, a major component of the collagen in ligaments and tendons. In about 12 percent of people, a molecule of adenine occurs at a point in the sequence where there's usually a cytosine (i.e., there's an A in the code instead of a C). This gene variant results in a form of collagen that makes ligaments and tendons thicker and stronger, less prone to rupture. And, perhaps since ligament tears are associated with a higher risk of osteoarthritis later in life, the mutant COL1A1 gene seems to confer protection against that, too. Kim told Brenner he had not just one but two copies of the protective allele, meaning he was among the less than 2 percent of people with an extremely small risk for an ACL tear, the most feared injury in football.

Just as the Bengals guard knew how much weight he'd been able to bench, Brenner, of course, knew that he had made it through 12 seasons in the NFL without any serious sprains. But he had always assumed it was mere luck. Caution about his future health had factored into his decision to retire in his early 30s. Every pro football player hears stories about guys who retire from the league and end up a few years later addicted to painkillers and hardly able to leave the house; he was afraid of being one of them. "I didn't know if I was going to be walking when I was fifty," he told Kim. Had he known he had less cause than most to worry about his joints, he said, he might well have played a few more years.

Talking to Brenner helped Kim see that injury risk, not speed or strength or broad-jump length, is the unknown elite athletes obsess over most because it's the one they feel least control over. "So much about being an elite athlete is not your performance —

it's not getting hurt," he says. "Who avoids injury determines your success as much as how good you are. But where you're going to get hurt, although your genes influence that, you don't really know that. Your VO_2 max, your heart rate, your leg strength — they know all that stuff exactly. What they can't see and what they don't know is: Where is the weak link? And if you knew that weak link, what could you do to keep the stress off it?"

Since his conversation with Brenner, while continuing to research the biological mechanisms of life span, Kim has opened a productive sideline as one of the world's leading researchers on the genetics of injury-proneness. When I spoke to him in the fall of 2016, his lab had just submitted four papers establishing links between specific alleles and various sports injuries. One paper, drawing on data from 100,000 people, found a marker that carries a 30 percent higher risk of rotator cuff injury for people who play sports with overhead arm motions, like throwing a baseball or spiking a volleyball. Rotator cuff tendinitis is among the most common injuries in baseball; major-league clubs spend millions of dollars each year on human experts and digital technologies advertising the ability to reduce injury risk to pitchers. If there were a wearable sensor or other technology that could reliably reduce pitchers' shoulder and elbow injuries by 30 percent, or even outperform chance in predicting such injuries, every team in baseball would be lining up to buy it.

Precisely because this sort of knowledge is so valuable, the pro sports world is being cautious in how it goes about incorporating genetic testing. A baseball team that knew its All-Star reliever had a genetic predisposition to rotator cuff tears could put him on a preventive strengthening program like the one Mackie Shilstone designed for Serena Williams and Peyton Manning. On the other hand, it could also use that information against him in contract negotiations, arguing that his services were less valuable than those of a hurler less likely to end up on the disabled list. For that reason, players' associations have been wary of genetic science. In many sports, unions have been reluctant even to embrace wearable sensors, worried the data they captured would be used in ways that would undermine athletes' negotiating power. DNA data, which reflects not just a player's current physiology or performance but his immutable destiny, is an order of magnitude more sensitive. Before genetic testing becomes commonplace, "we'd better think hard about the ethical and legal implications, especially for professional sports, where the contracts are so big," Kim says. "The collective bargaining agreement had better be in place so that the athletes are protected and they can't be discriminated against. Nobody wants to see *Gattaca* happen."

While big-money team sports figure out how to balance health and genetic privacy, Kim has been working with Stanford's cross-country team to explore how DNA data can help endurance athletes avoid injury. Running represents a considerably less fraught arena for Kim's work. For starters, the money's smaller, and professional runners generally compete as individuals so there's no team for whom genetic secrets might pose a conflict of interest. "So there's a little bit more of a clear path for ethical use of this information," Kim says.

For Stanford's cross-country runners, there's no injury more worrisome than stress fractures. A remarkable 40 percent of team members will suffer a stress fracture — a tiny bone fissure caused by overuse rather than acute trauma — over the course of a season. In theory, stress fractures should be preventable, since the principle of progressive overload dictates muscles and bones will respond to the stress of training by getting stronger as training volumes increase. In practice, distance runners typically *lose* bone mass throughout the season, for a web of interrelated reasons. One is energy balance: since excess weight means slower times, marathoners strive to stay ultralean and often fail to take in enough calories and nutrients to replace the ones they're using up. In particular, low levels of calcium in the blood cause bones to shed their calcium and become more porous. Endurance athletes, especially the ones who aren't eating enough, are also susceptible to overtraining syndrome, a condition in which accumulated fatigue disrupts endocrine function, causing the body to stop recovering and regenerating in between workouts. Elevated levels of the stress hormone cortisol, a feature of overtraining syndrome, further accelerates bone breakdown. In female athletes, all this is frequently compounded by amenorrhea, or the absence of a menstrual cycle. Anywhere from 20 to 60 percent of elite women endurance athletes experience disruption of their periods at extremely low levels of body fat. Just as osteoporosis becomes a serious risk for women after menopause, female athletes who don't menstruate experience faster loss of bone mineral. "There's this balance between overuse and performance," Kim says. "They would just keep on running more and more and more and get faster and faster and faster, except that they would all get hurt."

Genetic differences account for as much as 80 percent of the variation in bone-mineral density between individuals. If a cross-country coach knows which of his runners are genetically predisposed to strong bones and which are prone to fragile ones, he can design training programs in which the former maximize their mileage while the latter sacrifice some of the pounding for specific preventative conditioning programs. "For every single person, you want to train them right up to the breaking point, except you don't know where the breaking point is," Kim says. Knowing your genes doesn't tell you your breaking point, but it might at least tell you if yours is likely to be sooner or later than average. In a sport where 40 percent of competitors hit their breaking points each season, that's eminently actionable intelligence.

DNA data isn't usually this user-friendly. Because it affects so many people, not just athletes, the genetics of osteoporosis has been the target of dozens of government-funded studies comprising tens of thousands of people, making it relatively well under-stood. The trade-off for this wealth of evidence is that these are studies of civilians, not athletes, so the applicability to sports is un-certain. Kim's work with Stanford's runners is aimed at establish-ing that the same factors driving bone loss in menopausal grand-mothers also cause stress fractures in elite athletes, but for now, it's just a hypothesis — a "leap of faith," he calls it. "Generally, the science around sports medicine is a lot weaker than the science on disease," he says. If you want to bring large amounts of data to bear on questions of sports medicine, then, the quickest way is often to start with an analogy.

That analogy contains an implicit statement about the com-plicated relationship between sports and age. While the actuar-ial tables may show elite athletes live longer than everyone else, they also experience aging more acutely. Paradoxically, they can be both the healthiest humans alive and pictures of premature de-crepitude. The extreme stresses they pile onto their bodies often cause them to experience the debilities of old age — fragile bones, arthritic joints, dementia — decades ahead of schedule. "In a sense, athletes age a lot faster than everyone else — they age in forty years, not eighty years," says Kim. It's that dichotomy that makes them fascinating subjects for a biologist who studies longevity.

Which returns us to why I'd reached out to Kim in the first place: in hopes of getting a read on how my aging as a person was stacking up against my aging as an athlete. Had I, closing in on 40, pretty much come to the end of my prime years for performance, such as they had been? Or might I, like Hoby Brenner, have reason to think my allotted span might be somewhat longer than the stan-dard twoscore flat? Kim generously offered to interpret my geno-type report, which I downloaded from the 23andMe website, all 600,000 SNPs of it. At his instruction, I fed the text file contain-ing those 600,000 pairs of nucleotides into a website he'd built for the purpose. Then we convened a video chat to discuss the results.

We started with a big one: ACL rupture risk. A table on my report lists nine genes that have been tied to a greater or smaller likelihood of a torn ACL; most of them play a role in producing or maintaining the proteins that make up ligaments. A blue bar chart shows how I stack up against 700 other people who've been tested for the presence of SNPs identified as risk factors. "You're pretty good!" Kim notes. "Right in the middle." Significantly, at location rs180001, I have genotype AC. This SNP is part of the gene for the COL1A1 protein, and A is the rare, protective allele that results in the production of more collagen and a hardier variety of ligament. I don't have two A alleles like Hoby Brenner, but being in the 12 percent of people who have one is still, as Kim says, pretty good.

In fact, it's even a little better than that because COL1A1 is also a major component of tendons, so my rare-ish genotype comes with a somewhat lower risk of a ruptured Achilles, too. "It doesn't mean it's not going to happen, but it's a pretty measurable difference," Kim says. In the case of ACL tears, the people at one end of the bar chart are about twice as likely as those at the other end to sustain one, which is more noteworthy when you consider that some tears are the result of extreme trauma, like the low tackle that ended Marcus Elliott's high school football career, and no amount of extra collagen is going to keep them from happening.

I've always found it curious that I, despite being a rather ungraceful athlete who makes up for a lack of skill by hurling his body around recklessly, have never sustained the commonest injury in sports, a simple ankle sprain. How likely is it, I ask, that I have COL1A1 to thank for that? "It makes sense," Kim says. "If someone were to look at your ligaments, they might be thicker than an average ligament. You're pretty rare."

Next, we look at my genes for bone-mineral density. Once again, I'm more or less in the middle; Kim characterizes my risk of stress fracture as "low average." I'll take it. After that, we go over a laundry list of alleles — genes that influence how I process different nutrients like calcium, vitamin D, and magnesium, genes that control my testosterone production and blood pressure. I learn that I have one copy of the ACTN3 gene, the one that produces a protein found in fast-twitch muscle fibers. I seem to have unusually high levels of hemoglobin in my red blood cells — the highest of the 600 or so people whose DNA is in Kim's database — but the red blood cells themselves are unusually small. "You might have something interesting going on with the amount of red blood cells in your body and how much oxygen you can carry," Kim says, a bit cryptically. Overall, he tells me, I have a genetic signature somewhere between a power athlete's and an endurance athlete's, with elements of both but without the concentration of characteristics that would make me a likely candidate to be elite at either.

Then we get to the genes influencing life span. "Look at this!" Kim marvels. I don't come from long-lived stock by any means, so I wouldn't be shocked in the least to learn that all those homozygous SNPs are the genetic equivalent of snake eyes. After a few moments of confusion, however, to my surprise and relief, Kim determines that the matching pairs we're looking at are the desirable ones. I breathe a sigh of relief. My genes may doom me to mediocrity as an athlete, but at least I stand a better-than-average chance of being a mediocre athlete for many decades to come. I can live with that.

7

CONSISTENCY IS A TALENT

*The Psychology of Athletes Who
Get Better with Age*

When cyclist Catharine Pendrel had visualized the ride that would win her an Olympic medal in Rio, she gamed out every single scenario in her head. A muddy track, a delayed start, a leg cramp, a dropped chain — she had rehearsed her response to any contingency. At least, she thought she had. Then, mere seconds after the starting gun fired, she crashed. Then her gear shifters stopped working. Then she crashed again. Only a masochist could visualize a sequence of events so unlucky. But then, it takes more than a little masochism to win cross-country mountain bike races.

Ranked number 14 in the world at age 35, Pendrel had flown from her home in Kamloops, British Columbia, down to Brazil with realistic hopes of medaling and perhaps even winning gold for Team Canada. But she was determined not to let those hopes harden into expectations. She had learned better four years earlier, in London.

Pendrel had entered the 2012 Summer Games, her second

Olympiad, as the clear favorite in her event, simultaneously holding the titles of world champion and World Cup leader. That plus the experience of her Olympic debut in Beijing, where as a relative unknown she had surprised the field with a fourth-place finish, suggested that anything other than a podium appearance would be a disappointment.

Disappointment turned out to be an understatement. Pendrel went into the race feeling dogged by a low-level lethargy that had afflicted her for several weeks, but figured she would perk up once the adrenaline started flowing. That seemed to be the case as she shot out to an early lead. But midway through the 90-minute race, her stamina failed her. "My body just shut down on me," she told me. "It was just like, 'Whoa, power's gone. I can't make my legs go faster.'" She ended up placing ninth.

Analyzing what had gone wrong in the difficult days that followed, Pendrel came to understand she had let the pressure get to her. All those expectations had left her in a mental space where the fear of failure loomed larger than the opportunity for glory. The constant background hum of anxiety, she believed, had flooded her body with stress hormones that sapped her energy and undermined her recovery. "It's the challenge of going into the Olympics as a medal favorite — it just fatigues you in ways that you're not aware of," she says. "I let the seriousness of trying to win a medal steal the joy away from what I was doing."

When I meet Pendrel, five months before her shot at redemption in Rio, it's obvious she doesn't plan to let a lack of joy stand in the way this time. We rendezvous at a coffee shop on Bear Mountain, on Vancouver Island outside Victoria, where she is slated to ride in the inaugural event of the Canada Cup series the following morning. For Pendrel, who has already qualified for the Olympics with a top-five finish at the 2015 world championships, the race is mainly a chance to test her fitness and support the local cycling community. Sunny and relaxed, she shows up to our meeting with her blond hair in a ponytail and a splint on her right thumb. The weekend prior, she explains offhandedly, she slipped on wet rocks making a turn and fractured the distal phalanx. "It doesn't extend into the joint, so as long as I keep that joint immobilized, I'm good," she says. (This will not prove accurate. A few weeks after our meeting, a surgeon will determine that Pendrel's bone isn't knitting properly and insert a metal pin to stabilize it. The site of the pin will become infected, necessitating a second surgery.) The following morning, when I arrive to watch the race, the first person I see is Pendrel, who hails me from the saddle as she pedals past with a huge smile and a wave of her injured hand.

While Pendrel's accident hasn't affected her outlook, it has got her thinking. Her broken thumb is only the third injury of her career, following two broken clavicles. All three injuries happened after she turned 30. For the first time, she has started contemplating her age as something that might put a ceiling on her ability to compete. Objectively, Pendrel isn't at the end of the age curve for her sport. Her friend and rival Gunn-Rita Dahle Flesjä, a six-time world champion from Norway, is still ranked number 2 in the world at 42. The average age of the women in the Union Cycliste Internationale's top 10 is 32.6. But Dahle Flesjä and Pendrel were latecomers to the sport by the standard of elite athletics, having started racing in their early 20s.

Pendrel grew up riding horses, not bikes, in the tiny village of Harvey Station, New Brunswick, taking up cycling for fitness at her brother's urging only after she went away to university. For cyclists of her generation, that's a common arc. The first true mountain bikes were built in 1978, in Northern California, and it wasn't until 1996 that cross-country cycling became part of the Olympics. (Dahle Flesjä missed out on a bronze medal by 38 seconds in the inaugural race in Atlanta.) Just as the greater prize and sponsorship money in big-wave surfing has drawn younger competitors to what had been an older-skewing sport, the increased visibility of mountain biking has produced a bumper crop of youthful talent. "Now you have twenty-two-year-olds that are as good as forty-two-year-olds because a twenty-two-year-old may have been on a bike already for ten years," says Pendrel. In the months before Rio, Pendrel had in fact been hearing a lot of buzz about a pair of 20-somethings, Switzerland's Jolanda Neff and France's Pauline Ferrand-Prévot, who were regarded by aficionados as the up-and-comers to watch. "It had kind of gotten into my head a little bit," she admits. Ultimately, she told herself the age gap was neither here nor there; she was still capable of being the fastest woman in the world on any given race day and that was that. "You kind of have to block out all that chatter because you don't want to feel old when you're not," she says. "When it's a nonissue, you don't want to let it become one."

To keep others from making an issue of it, Pendrel has adopted a policy of strategic ambiguity around her plans for after Rio. Candidly, she's thinking it will be her last Olympiad. While her fitness is still world-class, her motivation has begun to ebb and flow from week to week. "To push yourself that hard, you have to really want it, and if you've been doing it for ten, fifteen years, no matter how old you are, I think it gets harder to do this," she says. But to head off any talk of retirement, she has committed to at least another season of racing and left the door open to more.

I met Pendrel through Trent Stellingwerff, who is supervising her Olympic preparations in his capacity as lead physiologist and head of research and innovation for the Canadian Sport Institute. Stellingwerff and I watch her Canada Cup race together from a muddy sward in the center of the racecourse. The 4.5-kilometer irregular loop begins near where we stand and twists up a knobby knoll, forcing riders to navigate a series of hairpin turns, then disappears into the trees along the flanks of Bear Mountain. After emerging from the woods, the trail descends sharply back to the sward, with part of the way transected by a rough staircase of wooden logs — one of the "features" designed to test riders' nerve as much as their skill. As we watch the riders complete lap after lap, their bright-colored spandex becoming progressively flecked with mud and sometimes blood, Stellingwerff explains that the physiology of a 90-minute cross-country MTB event like this is in some ways more similar to a soccer game than, say, a half marathon. In comparison with a road race, which is mostly steady-state spinning, all those dips and twists and jumps require the ability to generate sudden bursts of power. "There are periods that you need to be very explosive," he says. "Covering a breakaway or a sprint finish — that's primarily going to be the thing you lose when you age."

At coffee, Pendrel had told me losing explosive power wasn't something that affected her much — she had never had much to begin with. "It's funny being at the national team camp here because every girl on the team can absolutely crush me in a twenty-second start or in a sprint finish, but only one has ever beat me in a race," she says. Her success has been built on phenomenal aerobic capacity, the ability to grind out kilometer after kilometer without wearing down. "Even though I don't have that fast-twitch ability, my two-to-five-minute power is sufficient that it doesn't matter," she says. "I can kind of cover that weakness." In races, Pendrel's strategy is to negate her disadvantage by driving the pace throughout, gambling that her opponents will be too spent to overtake her in the final push. "I go out hard and try to make it so hard that nobody else has a good sprint left, either," she says with a laugh. It's essentially the same playbook Meb Keflezighi used to win the 2014 Boston Marathon, only Pendrel does it without the element of surprise, over and over again — just as she does on this cool March afternoon in Bear Mountain, where she shocks no one by finishing first.

And that's what makes what will happen in Rio five months later so remarkable. On a humid 90-degree day, Pendrel's hopes of dictating the race pace took a tumble when she did, amid a pile-up of bikes and bodies in the transition from the 600-meter opening loop onto the first 4.8-kilometer lap. "For those left behind, it's going to be a long, long haul," the NBC commentator said airily, surveying the fallen riders. Worse, after she got back onto her mount, Pendrel discovered the crash had jammed her derailleur. She was forced to climb Flag Mountain, the biggest hill on the course, with the chain on her large front sprocket. A quick visit with her mechanic in the tech zone fixed her shifting problem but left her in 25th place. For a moment, Pendrel felt herself becoming demoralized. Then her mind flashed back to her 2008 Olympic ride in Beijing, where she had surged from a similar position to finish in fourth place. She summoned up the previous May, when, at the World Cup in La Bresse, France, she shook off a terrible first lap and closed an 80-second gap to finish in second, behind Neff. She told herself that her number 1 goal coming into Rio had never been to bring home a shiny bauble. It was to ride a race she could be proud of. Even if every single rider in front of her rode a perfect race from that point on, her goal was still entirely within her power to achieve, she thought.

On her next trip up Flag Mountain, the height afforded Pendrel a clear view of the track ahead, and she was encouraged to see the leaders weren't as far up as she'd feared. Pouring it on, she began to pick off weaker riders, one by one, until she was in third place, a comfortable 25 seconds ahead of the next rider, who happened to be her countrywoman, Emily Batty. The 28-year-old Batty was as famous back home for her pinup bikini photo shoots in cycling magazines as she was for winning races. Pendrel was cruising toward the finish when, with 200 meters to go, she misjudged a jump and found herself drifting sideways in the air. Another crash. Pendrel scrambled to her feet, jumped onto her bike, and pedaled furiously for the finish, craning her neck to see Batty surging behind her. Pendrel crossed the finish line with a two-second margin. "If I just had another hundred metres I would have for sure got it,"

Batty grumbled to Canada's *National Post* — accurately, no doubt, given her superior high-end power. But that only made Pendrel's third-place medal all the sweeter to her. "I got everything I wanted out of that performance. It was far from perfect, but it was magic."

To be an elite athlete over a long period in any sport, from table tennis to triathlons, it takes rare physical attributes and consistent-enough health to make the most of them, both factors that are strongly influenced by genetics, as we've seen. The physiological demands of elite sport are so extreme and unyielding that professionals in the world of performance science often default — in case you haven't noticed by now — to the language of machines. I must have heard a dozen different versions of the "Ferrari engine with a Honda frame" analogy that Fusionetics creator Micheal Clark uses to characterize fit athletes who lack movement efficiency. A rusty Ferrari will usually beat a mint-condition Honda, former Red Sox general manager Ben Cherington has said. A Formula One car needs to be serviced more diligently than a Ford Focus, Trent Stellingwerff points out. If you had to drive one car your entire life, you'd care for it the way athletes care for their bodies, says Joel Sanders of Exos. The variations are endless, but the point is always the same: an athlete is only as good as her hardware.

But human beings aren't cars. The ones with the biggest motors and smoothest power trains don't always win, even in sports we think of as pure expressions of physics, like sprinting or power lifting. Unlike machines, humans have ideas, moods, emotions, skills, habits of mind. They can plan, strategize, visualize, and meditate. They can will themselves to moments of greatness, get into the zone or choke, intimidate or encourage one another. The older an athlete gets, the more these nonphysical attributes and experiences are the determinants of success or failure. By the same token, the greater an athlete's mastery of the mental and emotional side of the job, the longer his or her career is likely to be.

Yes, Catharine Pendrel's ripped legs and massive cardiovascular capacity helped get her to the starting line in Rio. But they're not what got her onto the podium. After all, many of the riders she was up against had the physical tools to medal. Look under the hood of Pendrel's performance and you'll find a collection of psychological and cognitive strategies that enable older athletes to achieve world-class performances long after their supposed physical peaks.

Begin with what she said about not letting anything "steal the joy away" from what she was doing. That was no throwaway remark. As far as she's concerned, joy made the difference between her subpar showing in London and the ride of her life in Rio. In the buildup to Brazil, "I was determined to remember that performing on the highest stage is fun," she says. "I always perform best when I'm smiling."

It's easy to hear sentiments like that as clichés, empty sound bites offered up by athletes media-trained into never saying anything interesting enough to cause a distraction. If you're any kind of sports-media consumer, you've probably heard that sort of thing hundreds if not thousands of times. For Green Bay Packers fans like me, it was a running joke during the 2000s that, at some point during a Packers game, one of the commentators would eventually offer the obligatory observation about how quarterback Brett Favre was "like a big kid out there" because he had so much darn fun, throwing snowballs at his teammates and the like. True to form, when I asked Favre what explained his unrivaled streak of 297 consecutive games started, that was exactly where his mind went. "I loved to play," he says. "There was numerous times I could've sat out, when I had a legitimate excuse not to play. When I broke the thumb on my throwing hand, I don't think anybody would've questioned me sitting out. The guy's got a broken thumb! But I wanted to play, loved to play, and I was determined enough to give it a try. You have to kind of start somewhere and I had that."

Stylistically, it would be hard to find a truer opposite for Favre —
a proud redneck whose second-favorite activity is sitting on a trac-
tor — than the elegant, cosmopolitan Roger Federer. But the Swiss
legend strikes all the same notes when talking about his own pro-
digious man-of-steel feats. In interview after interview since his
30th birthday, Federer has attributed his longevity to a deep and
abiding love of tennis. To hear him tell it, he, too, is just a big kid
out there. "I grew up playing against walls and cupboard and ga-
rage doors, and I still enjoy it," he said in an on-court interview at
the 2015 U.S. Open, where he had just beaten Stan Wawrinka in a
semifinal match. John McEnroe, the seven-time Grand Slam winner turned coach and broadcaster, says he has never met someone
who loves tennis as much as Federer — and McEnroe has probably met every great player of the last 40 years. Federer embraces
the aspects of his profession many other tennis players find most
grinding — the workouts, the daily practices, the constant travel.
He has even claimed that rehabbing his damaged knee was a good
time, if only because any sort of injury was such a novelty to him.
"I look at these tougher times as something super interesting and
almost, in some ways, something fun," he said after getting his meniscus repaired.

Well, duh, you're thinking. *The people who spend their lives playing games and going to the gym actually enjoy it.* But to sports psychologists, there's nothing obvious or trite about the way people
like Federer, Favre, and Pendrel talk about their jobs. Quite a lot
of professional athletes don't think that way; we just don't hear as
much about them because they're the ones who don't stick around
nearly as long and don't perform as well while they last. "Fun is really *the* critical element," says Jim Loehr, cofounder of the Johnson
& Johnson Human Performance Institute. In more than 30 years
as a sports psychologist, Loehr has counseled the likes of tennis
stars Jim Courier and Monica Seles, hockey player Eric Lindros

and speed skater Dan Jansen. His work has bred a conviction that a consistently high level of joy is the biggest driver of long-term athletic success. "Fun is probably the best mirror you can look at," he says. "I'm always looking at how much fun does an athlete have at practice, how much fun does he have in the gym, how much fun does he have in moving to the next level."

If a healthy sense of fun is surprisingly rare, it's likely because one of the first filters that separates the youth athletes who become elite performers from those who don't is the self-discipline to choose a life of grim sacrifice over one of easy gratification — to put in those 10,000 hours of practice required to achieve mastery (or however many it takes), to eat and sleep in a regimented way, to break down every performance afterward and attempt to understand everything one did wrong. In the life of an adolescent gymnast or swimmer hoping to make an Olympic team, there's little room for playfulness or spontaneity. But if the ability to treat a game like a round-the-clock job is crucial to early success, the ability to nurture a spirit of delight despite all the forces threatening to extinguish it is equally vital later in one's career. "We have this puritan work ethic — 'work hard, work hard, work hard.' That can kill our joy," says Jim Afremow, a Phoenix-based sports psychologist who has worked extensively with professional and Olympic athletes. "The best players in the world work the hardest, but they also have the most fun."

After winning his fifth Super Bowl, Tom Brady explained his success this way: "Other than playing football, the other thing I love to do is prepare to play football." Conceptualizing all that hard work as something to be relished in itself rather than as a means to an end is an underappreciated cognitive skill of top athletes. Those who fail to develop that skill — who think of weight-room sessions or mile repeats or back-to-back road games as the price to be paid en route to an ultimate reward — may reach the highest level of their sport but they're unlikely to stay there. Eventually, they will suffer a decline in what psychologists call intrinsic motivation, which basically just means enjoyment in doing.

"If you're playing solely to rack up titles and get as much money as you can, it's kind of joyless," says Loehr. "But if you're doing this because it's a vehicle for expressing who you are, with a sense of gratefulness and joy that pervades both winning and losing — if I sense that in an athlete, I don't care what the sport, you know they're going to hang around a very long time."

"The research on intrinsic motivation is really robust," Afremow agrees. In his clinical practice, he often encounters slumping players who say a prolonged run of bad performance has made them miserable. "My response is: 'Why would you be playing well if you're not having fun?'" he says. Afremow's *beau ideal* of intrinsic motivation is Michael Jordan, who came out of retirement not once but twice and played until he was 40. Jordan famously competed as hard in practice as he did in games, preferring to team up with second-stringers for scrimmages for more of a challenge. It's no coincidence, Afremow says, that Jordan was also the first NBA player to insist on a "love of the game" contract clause granting him the right to play pickup ball in the off-season. Jordan is probably one of the few players who ever loved basketball as much as Federer loves tennis. "All successful performers love what they do," Afremow says.

But there's more to it than "love what you do." There's no such thing as an elite athlete who's 100 percent internally motivated. Every great competitor needs measuring sticks. That's where goals come in.

Defining one's goals can be a risky proposition. An ambitious goal can inspire an athlete to a new level of performance, but it can also have the opposite effect if it goes unmet or is perceived as too intimidating. The importance of careful goal selection is magnified in older competitors experiencing age-related physical decline. For athletes used to being able to perform at a certain level, the discovery that they can no longer do so can undermine not only motivation but even their sense of identity, says Don Kalkstein, the head of performance psychology for the Dallas Mavericks. As a result, the best older athletes tend to develop a sophisticated and nuanced approach to goal setting, one that marries ambition to realism and provides for multiple paths to fulfillment.

One thing such athletes pretty much never do is commit themselves to one specific vision of success. When Pendrel fell behind on the first lap, the first thing she did was remind herself that she didn't need to attain any particular finish place or time to have a good Olympics. That attitude was a product of her humbling experience in 2012. Having gone into London as the presumptive gold medalist, "I had to be perfect, and anything less wasn't enough. That unforgiving mind-set gives you no room to come back from adversity," she says. "It means a solid performance can quickly derail into a poor one."

Pendrel's statement is an example of what psychologists and coaches call being process-oriented rather than outcome-oriented. The problem with focusing on outcomes is they're never fully within your control; even if you run the best race of your season, the guy in Lane 2 might choose that day to run the best race of his life. Learning to separate process from outcome is a critical skill for athletes, says Jack Groppel, who cofounded the Human Performance Institute with Jim Loehr. "I tell people: 'When you're in control, be brutally honest with yourself. When you're not in control, be gentle with yourself.'"

Teaching people how to keep unexpected setbacks from turning into catastrophic spirals was a major focus for a clinical psychologist named Eric Potterat during his 20 years in the U.S. Navy. Among Potterat's responsibilities was designing the curriculum for the Survival, Evasion, Resistance, and Escape course, the grueling program Navy SEALs and other elite special operators must take to prepare themselves for the worst contingencies, like being captured and tortured or stranded in the desert. Later, as the first head psychologist for the SEALs — the secretive amphibious force whose exploits include killing Osama bin Laden and rescuing the crew of the *Maersk Alabama* from pirates — Potterat developed something called the Mental Toughness Training Program.

Lecturing the world's hardiest warriors on the finer points of toughness is an undertaking that itself requires no small amount of intestinal fortitude. Just to get into the SEALs, candidates endure something called Hell Week, whose trials include extreme sleep deprivation, sitting for hours in icy water, and thousands of push-ups. Imagine doing all that and then hearing from some *psychologist* about toughness. Sparely built and of medium height, Potterat is not going to be mistaken for one of the burly, bearded killers from *Zero Dark Thirty,* but he has the kind of quiet authority that makes you lean in and listen to whatever he has to say. Since retiring from the Navy in 2016, he has turned his attention to athletes, joining the Los Angeles Dodgers as director of specialized performance programs. I met him at the Los Angeles headquarters of Red Bull, where he was leading a workshop in performance under pressure for a group of the energy drink's sponsored athletes. Several of the attendees, including backcountry skier Michelle Parker and motorcycle racer Aaron Colton, competed in sports where fatal accidents were not uncommon; their interest in how SEALs stay cool in life-or-death situations was more than academic. They scribbled notes as Potterat drew a large circle on a dry-erase board and then asked the athletes to name the things that cause them to feel pressure during a performance. Potterat wrote their responses — "noise," "other people's opinions," "other people's mistakes" — outside the perimeter. Inside it, he wrote, "Things I Control: 1. My attitude. 2. My effort. 3. My actions." When a performance — or a mission — begins to go pear-shaped, the key to preventing it from snowballing into a true disaster is to focus on the things you can control and *only* on those things, he said. The mantra he gave SEALs to repeat when things got hairy: "Stay in your circle."

Pendrel stayed in her circle when she remained calm and upbeat during what could have been a ruinous first lap. But she also did something just as important: she recalibrated her goals on the fly to ensure she would feel challenged but not discouraged. Having started the race with one notion of what success would look like — leading from the front, choosing the pace, avoiding a last-minute dash — she set it aside in favor of another one that was more realistic given her changed circumstances. In effect, when the carrot she was chasing got too far away, she had another carrot at the ready.

For a textbook example of using flexible goals to stay motivated in the face of adversity, look at Meb's experience at the London Olympics. Going into every race, Meb said, he had a set hierarchy of objectives: first, to win; failing that, to earn a place on the podium as one of the top 3 finishers; failing that, to finish in the top 10 or achieve a personal record. But midway through the 2012 marathon, even the attainable goal was looking out of reach. He had arrived in London in less than peak fitness, having battled through injuries in the months prior. At an early aid station, Meb reached for his water bottle only to be handed one belonging to his teammate, Ryan Hall. Elite American marathoners hydrate with sports drinks customized to replenish their carbohydrates and electrolytes in the exact proportions they lose them. Meb resisted drinking from Hall's bottle at first, worried it would upset his stomach. Finally, concerned about dehydration, he gave in and drank, only to be proven right. Wracked with stomach cramps and a blister on his left foot, he fell back to 21st place. He considered dropping out to spare his body for the upcoming New York City Marathon.

But he didn't. Instead, he focused his energy on the humblest goal in front of him: to finish the race. As he began to rally, he traded up to a slightly more ambitious goal: to hang with the Japanese runner Kentaro Nakamoto, who was making a surge. Riding

Nakamoto's tail, he broke into the top 10. At that point, he could have backed off, secure in the knowledge that he'd achieved at least one of his prerace hopes. But he was thinking about a conversation with his coach, Bob Larsen, who had urged Meb to fight for every single place in the top 10. The World Anti-Doping Agency had recently begun freezing samples of athletes' blood and urine for future analysis, to allow them to be tested with not-yet-discovered methods. Some runners from the 2004 Athens games had just been stripped of their medals when frozen specimens revealed banned substances that had gone undetected the first time around. Even a fifth- or sixth-place finish might eventually turn out to be worth a medal, Larsen had pointed out. With three miles to go, Meb caught a glimpse of his coach standing on a raised platform, holding up six fingers at him and shouting. Meb was in sixth place. Suffering as badly as he ever had, he expended everything he had left over the final 15 minutes to place fourth. Unlike Emily Batty, Meb was ecstatic to finish just outside the medals. To this day, he considers it his finest effort. So does Larsen. "I've seen Meb just produce something out of nothing before," he told me, "but that was pretty spectacular." (And it will be even more spectacular if one of the three finishers ahead of him gets busted on some future doping test.)

To say older athletes must learn to reorient their thinking around realistic goals when their original ones appear unattainable might sound suspiciously like saying they ought to learn to settle for less — like they should curb their ambitions as their bodies decay. In fact, it's exactly the reverse. Setting multiple tiers of goals and continually reassessing them as circumstances shift is all about never settling for less than is possible *in that moment*. It's about not using perfectionism as an excuse to quit when the glory of a trophy or a medal is out of reach. Or maybe it is about embracing perfectionism, but perfectionism of effort, not outcome: stay-in-your-circle perfection. It's precisely the mind-set you would hope to find in an older competitor who approaches sport with a sense of joy and intrinsic reward: a reluctance to abandon any effort, however jinxed, because every effort is a privilege and who knows how many more you'll be granted?

The concept of using flexible or relative goals to motivate is codified in masters sports, where everything is about age-group competition and age-adjusted records. Using standard tables maintained by the World Masters Athletics governing body, athletes are able to compare their results against those of competitors in different age brackets much as a high handicap allows a duffer to play a competitive round with a scratch golfer. When I plugged my time of 23 minutes for a recent 5K run into the WMA calculator, for instance, it shaved off more than a minute, meaning that's how much faster I would have been expected to run in my "prime" of 18 to 29. Age adjustment means an athlete can get "faster" just by maintaining the same times year after year, or even by slowing down at a slower-than-expected rate. From a psychological standpoint, that's huge. David Walters, an airline pilot and onetime Olympic marathon qualifier, has used WMA's formula to challenge himself in the 25 years since he ran his lifetime personal best of 2 minutes, 19 seconds at the 1988 Olympic trials. At the 2015 Chicago Marathon, he finished in 2:45:26, which was not only the fastest time for a 60-year-old by more than 20 minutes but also graded out as 3 minutes faster than his 1988 run after age-adjustment. "At this stage, when you're working your way backward in time, you have to find new motivation," he said afterward. "You know next year you probably won't be as fast, but you still find the joy of pushing yourself to the limit." (You might be using age-adjusted goals without realizing it. *Not bad for a guy who's almost forty*, I'll often say to myself after chasing down some youngster on the soccer field. Who knew my overgenerous self-regard was a legit motivational tool?)

Goal setting plays a significant role in one of the biggest concrete advantages older athletes have over the young: superior mastery of emotions. Andy Lane, a sport psychologist who studies the impact of mood on performance at the University of Wolverhampton in England, says young athletes often suffer from intense emotions like anxiety and anger that affect them before and during performances. These emotions aren't just unpleasant; they're also disadvantageous in a number of ways. For one thing, athletes who experience unwanted feelings often expend significant time and energy on strategies to combat them. These can range from mundane ones, like meditation, to bizarre ones, like the dozen or so superstitious rituals Rafael Nadal goes through on match days. There's even data suggesting that unwanted emotions consume actual physiological resources. In a study published in the journal *Applied Psychophysiology and Biofeedback,* researchers had competitive cyclists perform the equivalent of a 10-mile time trial on stationary bikes while wearing masks that collected their exhalations. Some were shown feedback accurately reflecting their performance; others were shown false feedback designed to distress them by making it seem like they were performing badly. While there was no difference in power generation or time to completion between the two groups, subjects in the false-feedback group consumed more oxygen and produced more lactic acid. The researchers concluded that positive emotions are associated with a reduced metabolic cost of performance.

The deleterious effects of negative emotions are even more readily apparent when it comes to skilled performance. Stress and anxiety manifest in excess muscle tension that can sabotage the coordination required for complex actions. "That's the kiss of death in sports," says Afremow. It's particularly a problem for so-called closed skills — those in which the athlete is initiating an action in a controlled, predictable environment, not reacting during the flow of play. A golf stroke, a tennis serve, a basketball free-throw, and a soccer penalty kick are all closed skills; an athlete performing them has time to think and plan, which also means time to experience unwanted emotions. When that happens, a sudden breakdown in coordination — a choke — is often the result. Golfers talk about a mysterious phenomenon called "the yips," a sudden shudder or spasm that strikes while putting and can send a 2-foot putt rolling 20 feet past the hole. Most attempts to explain the yips focus on arousal, the physiological state of excitement or anxiety. A little arousal is good, bringing heightened concentration and awareness, but too much and the risk of choking goes up.

The good news is that athletes tend to experience fewer and less-intense unpleasant emotions as they get older, according to Lane, whose research focuses on mechanisms of emotional regulation. While not universal, he says, athletes who continue to perform at a high level over many years typically become better at avoiding "unwanted psychological states" before and during competition. "I am not sure whether older people have fewer unwanted emotions in the first place, or whether they regulate them very early in the process and so take very early action," he told me. "We need detailed studies that look at the speed of regulation and prevention of intense emotions — difficult to do because you need to know that the person actually prevented an intense emotion occurring." Strategies that can defuse unpleasant emotions include self-distraction, positive self-talk, and cognitive reassessment. Self-talk, in particular, has shown itself in study after study to be a powerful tool with surprisingly concrete benefits. In one 2013 study, cyclists who received two weeks of training in self-talk, learning to identify negative thoughts and replace them with positive ones, were able to go 18 percent longer on a time-to-exhaustion test than members of a control group.

But there's no cure like prevention, and Lane believes a difference in goal-setting style is what helps at least some older athletes avoid finding themselves behind the emotional eight-ball to begin with. In large-scale studies with runners and triathletes, he has seen that young athletes tend to identify challenging goals such as setting a new personal best, even in conditions — say, a rainy or hilly course — where accomplishing that is unlikely. Older athletes are better at setting realistic goals. That's not surprising, Lane notes: "Older people, of course, have had more experience of attempting to achieve and thus expectations are modified by experience." But it is important, since failure — particularly when coupled with a misplaced sense of control that deepens one's feelings of having failed — can have all kinds of downstream consequences. In the short term, excessive feelings of failure can create the kind of anxiety or agitation that raises the likelihood of choking. In the medium term, they can create stress or depression that interferes with sleep and recovery, leading to more poor performances. In the worst cases, they can shorten careers, says Loehr. "Some athletes will go days where they won't eat, won't sleep. They feel they have to punish themselves for playing badly," he says. "Those athletes don't last long."

In no other sport is the importance of emotional regulation more obvious than in tennis, with its blend of physical exertion, isolation, and pressure situations. Start with the raw endurance factor: it's not unusual for a men's Grand Slam match to last four hours, twice as long as a marathon. In the course of that match, each player will likely have to serve — i.e., perform a complex closed skill — somewhere between 100 and 200 times. By comparison, in golf, sometimes considered to be the most psychologically challenging sport, the average player on the men's tour completes a round in about 70 strokes.

And golfers have their caddies to talk to. Doubles matches aside, the tennis player competes in a state of unrivaled existential solitude. In his memoir, *Open*, Andre Agassi put it this way:

Only boxers can understand the loneliness of tennis players — and yet boxers have their corner men and managers. Even a boxer's opponent provides a kind of companionship, someone he can grapple with and grunt at. In tennis you stand face-to-face with the enemy, trade blows with him, but never touch him or talk to him, or anyone else. The rules forbid a tennis player from even talking to his coach while on the court. People sometimes mention the track-and-field runner as a comparably lonely figure, but I have to laugh. At least the runner can feel and smell his opponents. They're inches away. In tennis you're on an island. Of all the games men and women play, tennis is the closest to solitary confinement.

And a tennis court, like a jail cell, is an easy place to lose your mind. The sport's upper ranks are filled with examples of players who reached the world top 10 only after struggling with epic on-court meltdowns in their teens or twenties. Federer, Andy Murray, Novak Djokovic, Serena Williams, and Agassi are just a few of the players who started out as brilliant talents prone to throwing tantrums or giving up when line calls went against them or a shot wasn't working. Then there's Stan Wawrinka. Unlike the aforementioned players, who vanquished their demons around the time their physical and technical development allowed them to start winning Grand Slam titles, the big-hitting Swiss continued to suffer from debilitating emotional collapses for a full decade after turning pro. He was known for three things: being one of the fittest men on the tour, possessing possibly the world's most beautiful backhand, and being unable to hold it together in high-pressure situations. When he finally got a handle on his psyche, there was no mistaking the impact on his success.

Wawrinka's turning point came in 2013, when he started working with a new coach, Magnus Norman. A former world number 2 himself, after retiring, Norman founded a tennis academy called Good to Great, and that was the effect he had on Wawrinka's game, mainly by helping him find ways to calm and refocus himself during matches. Before Norman, Wawrinka was seen, as his coach would later say, "a little bit nervous and a little bit soft when it came to tough matches." Specifically, Wawrinka was prone to two types of mental lapses: he had a habit of "pressing," or trying to do too much, in key moments, and beating himself up when he felt like he had blown an opportunity. Both were linked to excess arousal. Norman worked with Wawrinka to level out his arousal level, making sure he never got too keyed up or too down in the dumps. The results were immediate: Wawrinka made it to his first Grand Slam semifinal in 2013. He was 28, an age when most professional tennis players have long since peaked. (Of being 28, Agassi wrote, "Every other article contains the threadbare phrase, 'At an age when most of his peers are thinking about retiring . . .'") The following year, he won his first major, and then he repeated the feat in each of the next two years — in the process, becoming only the third man in tennis history to win multiple Grand Slam championships after age 30.

Wawrinka's most dramatic moment came at the 2016 U.S. Open, where he played Djokovic, the reigning champion and world number 1, in the final. Five minutes before taking the court, Wawrinka sat in the locker room, shaking and weeping with nerves. Things only got worse once the match started. After falling behind in the first set, Wawrinka pushed Djokovic to a tiebreak, then won only one point in it. It was a classic blown opportunity, Wawrinka's emotional Kryptonite. He promptly spotted Djokovic a 4–1 advantage in the second set as well. The most consistent player in the game, Djokovic was also fresh, having benefited from his opponent's retirement in three of his six matches leading up to the final. That helped him to play his preferred style of match, heavy on defense and long, grinding baseline rallies. It looked hopeless, but Wawrinka found the unlikeliest of edges in his own exhaustion. As Djokovic ran him back and forth, Wawrinka burned off the nervous energy that had been plaguing him and settled down.

Once his attention shifted from how his mind felt to how his legs felt, he started winning points. After each big one, he locked eyes with Norman, seated in his supporters' box courtside, and silently pointed to his temple. The gesture, he said afterward, was a form of self-talk, a reminder to himself to stay focused on the next point, not the last one. By the end, it was Djokovic who was coming apart, violating the rules to take a dodgy injury time-out that only delayed the inevitable.

In the postmatch press conference, Djokovic was asked what had happened. His answer showed how completely transformed Wawrinka was from the days, just three years earlier, when he had been regarded as a gifted technician and physical specimen who melted under pressure. "He plays best in the big matches," Djokovic said.

It's easy to think about the traits we associate with youth — passion, optimism, a sense of invulnerability — and assume they translate into psychological advantages in sports. And no doubt they do sometimes. But just as often they cut the other way. Passion boils over into racquet-smashing fits of rage or crippling performance anxiety. Optimism and self-belief cause athletes to expect too much of themselves too soon and blame themselves for not achieving it, or prepare less than they should. Without a doubt, there's something to be said for youthful enthusiasm and resiliency. But most of us overrate the value of those characteristics and underrate the importance of the ones that accrue with age. Maybe it's because of our society's obsession with staying young, or the marketing messages that go along with it. Maybe it's because we see the young dominating in so many sports and attribute it to the wrong factors, assuming they're succeeding because of all their youthful qualities rather than in spite of some of them. Maybe it's because believing in the primacy of things like passion, desire, and single-mindedness has more romantic appeal than the reality.

When you watch the promos during a televised tennis tournament, it's all players pumping their fists and shouting *"Come on!"* or howling at the sky. The implication is it's a battle of who wants it more. But Stan Wawrinka wasn't underachieving in big tournaments because he didn't want it enough. If anything, his passion burned too hot. He had to learn to bank it, keep it at a simmer. Wawrinka needed to temper his passion with consistency, and consistency is exactly the kind of boring-but-vital quality we associate with older competitors. It was consistency, not passion, that allowed Meb to be out front in Boston on a day when the fastest marathoners in the world let their guard down just a little too long. "A lot of athletes do everything right and you think they're ready to go, but then they go to the race and they're just flat," Bob Larsen said of Meb as a competitor. "Meb, when he's healthy, he'll at least be close. Maybe that's part of your talent, getting the most out of yourself in big competitions."

Consistency, a knack for methodical planning and goal setting, a realistic knowledge of one's capabilities, and a healthy sense of perspective — none of these sound very sexy. "Set challenging but attainable goals" is not going to replace "Impossible is nothing" or "Just do it" in running-shoe ad campaigns anytime soon. But they are the qualities that, more often than not, determine long-term athletic success and help older athletes overachieve. If we don't appreciate this as much as we should, it's because they often don't fully manifest until other, more obvious athletic traits, like speed and agility, have started to wane.

But they're getting harder to overlook as older athletes are having ever more success, and nowhere is that happening more than in tennis. In 2006 the average age of players who won an Association of Tennis Professionals event was 24. By 2016, it had jumped to nearly 29. Over that decade, the number of tournament champions over 30 increased from a round 0 to 14. While Wawrinka may be an outlier in starting to win Grand Slams after 28, the ranks of his peers now teem with 30-something players who say their maturity allows them to win matches they would have blown in their 20s. "It's not that you're any more talented at that age, it's just that you've figured out how to get the most out of your talent," James Blake, an American who played until 33, told the ATP's website (in an article titled "Why 30 Is the New 20 on the ATP World Tour"). "You know how to handle the pressure." "You're a little calmer in the face of adversity," former world number 1 Andy Roddick agreed. Sound familiar?

8

SLOW IS FAST

Strategy, Complexity, and the
Advantages of Experience

James Galanis had a problem with his protégée: the best woman soccer player in America was too fast.

It was 2012 and Galanis, an Australian soccer coach who runs a youth clinic in central New Jersey, had been working with Carli Lloyd for nearly a decade, since the day her father, Steve, had approached him on the sideline after a practice to say his daughter needed help. Carli, then 20, had just been cut from the U.S. national under-21 team, told her game had big flaws that would need mending if she wanted a shot at the senior team. Devastated, she had decided she'd rather quit soccer and get a civilian job than try to eke out a living at the sport's lower levels, which, for women at that time, were all but nonexistent anyway. Her father had convinced her to give it one more shot and set up an evaluation with Galanis, who had a reputation for molding local talents into pros.

At their first meeting, Galanis put her through a number of drills and saw the under-21 coach had not been wrong to drop her. Lloyd had the natural touch of someone who'd grown up dominating her club matches and street pickups, but her technique was raw, even when it came to basic skills like receiving and controlling passes or dribbling at speed. Worse, her fitness was pathetic. Just running drills exhausted her. When Galanis had her run laps to test her endurance, she had to stop after 12 minutes. Worst of all, she didn't seem to believe any of this was her responsibility. Whenever Galanis pointed out something she was doing wrong, Lloyd, who had played on a Division I team at Rutgers University, found a way to blame it on her coaches or teammates. Before he could do anything about her technique or fitness, Galanis thought, he would have to fix her character. "That's where I'm different," says Galanis. "I build my players from the brain down, not from the legs up. I make sure they're coachable."

Galanis told her he would mentor her, free of charge — but only on the condition that the excuses stop. If they were going to work together, her commitment would have to be total. "A regular person works nine to five, okay? And at five o'clock, they shut down. They have beers. They do whatever it is they want," he told her. "The rest of the world has an off switch that happens every day, but there's no off switch for you. Your off switch comes at the end of your career. And that might be at thirty-five. That might be at thirty-six. From this point on, the switch is on and it doesn't come off until you take off your boots and you've retired from the game."

Lloyd, who had grown up cosseted by coaches who were just thrilled to have such a talent to help them win games, was galvanized by Galanis's challenge. She agreed to his conditions. After they began working together daily, Galanis realized the task was even bigger than he'd appreciated. Lloyd's poor performance on his fitness tests was only partially a result of laziness, he discovered. She simply had low baseline cardiovascular fitness. If she wanted to play international soccer, she would likely always need to do substantially more conditioning work than her teammates. Galanis put her on a regimen that included 90 minutes of running,

400 push-ups, and 1,000 sit-ups — every day. Even at that volume, it took almost two years for her to reach the national team's fitness standards.

But the job was bigger in another way, too. The more he saw of Lloyd, from her inner drive to improve to her ability to summon genius with the ball at her feet, the more Galanis came to believe she had it in her to be a once-in-a-generation player. Maybe more. Mentor and pupil drew up something that sounded crazy: a three-stage, 10-year plan to become the best soccer player in the world. "Even my wife didn't believe me," he recalls. "She's looking at me saying, 'This bloke's fucking messed up. He found this girl here in Delran, New Jersey, and he thinks he's going to make her the best in the world.'"

That's pretty much exactly how it went down.

The goal of Phase 1, which lasted from 2003 to 2008, was what Galanis called "Get your foot in the door." He and Lloyd ironed out the kinks in her technique, like an inefficient way of receiving passes that produced sloppy giveaways, and brought her fitness up to par. She established herself as a hard worker on the field as well, erasing a reputation for being a ball hog who slacked on defense, although she remained prone to mental mistakes. Lloyd quickly won back her place on the U-21 squad. In 2005, six days before her 23rd birthday, she made her first appearance for the senior team.

Phase 2 started ahead of schedule. The objective was to attain recognition as the best American woman — but in fact Lloyd had already arguably done so at the end of 2008 when she was named U.S. Soccer Athlete of the Year in voting overseen by the U.S. Soccer Federation. (Goalkeeper Tim Howard was her male counterpart for the honor.) It capped a year in which Lloyd led Team USA to an Olympic gold medal, scoring the game winner in the final against Brazil, and started every national team match. Even Galanis's wife had to admit he was onto something.

As auspiciously as Phase 2 began, Phase 3 was just the opposite. Gearing up for the Olympics in London, Lloyd got a devastating piece of news from Pia Sundhage, the USWNT coach. She had been demoted to the bench. Lauren Holiday and Shannon Boxx would be starting in central midfield, Lloyd's position. Sundhage thought they would be more careful with the ball than Lloyd, who still had a reputation for taking unnecessary chances. Lloyd also suspected that Sundhage still blamed her for the team's loss at the World Cup the previous summer, where she was one of three players who missed penalty kicks during an overtime shootout in the final against Japan.

Lloyd's career as a reserve lasted exactly 16 minutes, until Boxx left the opening match against France with an injury. Lloyd proceeded to score the game winner and lead Team USA to the gold-medal match, a replay of the previous year's World Cup final — only this time Lloyd didn't waste her chances, scoring both goals in a 2–1 victory. After the tournament, Lloyd confessed she was driven in part by a desire to prove her coach wrong. She got her wish and then some: Sundhage resigned as coach shortly afterward and moved back to Sweden.

In writing off her own best player, Sundhage was gambling that a midfielder who had just turned 30 and had never been all that impressive a physical specimen to begin with had passed her sell-by date. In fact, Sundhage had said as much after pulling Lloyd from a friendly against China in May, a few weeks before the Olympics. "After that game, the coach came and said, 'We think you've slowed down. You're not considered a starter anymore,'" Galanis says.

What Sundhage didn't reckon on was that slowing down was exactly what Lloyd and Galanis were working on. That was Phase 3. For a decade, Galanis's master plan to make Lloyd the undisputed best female soccer player had been building up to it. After playing on the international level for seven years, he reckoned, she was finally ready.

On a chilly spring morning, I drove the 90 minutes from Manhattan to Mount Laurel, New Jersey, where Galanis's Universal Soccer Academy is based. I parked on a quaint main street full of picturesque Federal-style homes. Galanis, who looks a bit like the comedian Jack Black, met me at the door of his office with a hearty "Over here, mate!" Inside, amid memorabilia from her career, he walked me through Lloyd's evolution and the latest stage of their decade-long program, occasionally demonstrating one of his points by kicking a junior-sized ball against the baseboards.

The major difference between NCAA soccer, American professional league soccer, and international soccer, Galanis explained, is the speed of the game. As you move up the levels, "everything happens quicker. You get shut down quicker. You get into space quicker. You have to do everything quicker because the game is so much faster." When college players like Lloyd first begin playing national team games, they're almost always stymied by the speed of play. Most react by running as fast as they can to find passing channels or close down opposing players. But physical speed is the wrong solution to the problem.

"If you want to play quicker, you can start running faster. But it's the ball that decides the speed of the game," the great Dutch player and coach Johan Cruyff once said. Not even the fastest player can outrun the ball. Being able to chase down the other team's lightning-fast winger isn't much use if, the moment you arrive, he launches a perfect cross to his striker in front of the net. The ability to outsprint any defender isn't much of an advantage if you run the wrong way. "The fastest player on the field isn't the person with the fastest body, but with the fastest mind," Galanis says.

And the fastest minds seldom belong to the youngest players. True, they have the fastest *reflexes*. Every neuron is wrapped in a protective protein sheath that acts as insulation, improving impulse conduction. As nerve cells age, this protein sheath degrades, leading to an increase in reaction times beginning in one's 20s.

But reaction time and decision speed are two different things.

In sports like soccer and other complex tasks, there's too much going on for the brain to process it efficiently all at once. Instead, our brains engage in something called "chunking," linking together related pieces of information into patterns to reduce the number of variables at play and streamline the decision process. A novice baseball batter might look at a defensive arrangement and see that the shortstop, second baseman, and first baseman are all playing closer to third base than they usually do; a more experienced one might note the shortstop shading over, realize the defense is playing a shift, and know to expect a right-handed pitcher to throw an inside fastball.

The Canadian researcher Janet Starkes was among the first to identify how chunking works in elite athletes. She did this by taking volleyball and field hockey players and showing them slides of game situations. Starkes flashed the slides in front of their eyes for a few hundredths of a second and tested their recall afterward. Novice players, she found, could barely discern anything, but expert players were able to remember even tiny details. The experts were able to process so much more information because they were reading the chunks, the same way recognizing words allows you to read this sentence in a few seconds while a child of five would be stuck sounding out the letters. An Australian researcher named Bruce Abernethy built on Starkes's work with "occlusion" studies in which novice and expert athletes were asked to describe situations or perform tasks in their sport while being denied critical visual information. A tennis player, for instance, might be asked to predict the direction of an opponent's serve after being shown her service wind-up but not the moment of ball contact. Like Starkes, Abernethy found a gulf between those who had spent years training and competing at the highest levels of their sports and those who hadn't.

When young players make a sudden jump in performance after three or four years as a professional, they almost always explain it using the same phrase: "The game slowed down." When an athlete makes a play involving an amazing act of anticipation — a cornerback undercutting a wide receiver's route for a pick-six, for instance — what you often hear is, "I saw what was going to happen before it happened" or, "It felt like it happened in slow motion." Chunking is the mechanism responsible for those subjective phenomena. It's what makes complicated things feel simple and deliberate actions feel automatic.

"When you're young you're fast and explosive but you don't know where you're going," says Galanis. Because the game *feels* so fast, young players feel like they need to *move* fast. The result is a lot of wasted, inefficient effort. It's only as they acquire a vast, chunky mental database of game situations that they're able to process fast enough to see the field clearly and respond to what's about to happen rather than what just happened. Once that occurs, "you're not running frantically trying to find space anymore. You've figured it out. You can almost walk into space."

One of the things that make soccer the world's most popular sport is how little you need to play it: a ball, a flat area, a few objects to mark the goals. This is true, to a degree, when it comes to athletic requirements as well. Compared with other major sports that demand specific body types, soccer is relatively catholic in the physiques on display. Zlatan Ibrahimovic stands nearly a foot taller than Lionel Messi, yet the two both play forward and are among the best in the world at it. The same goes for speed. Recall that even though it's nominally a game of sprints, soccer players are somewhat more likely to have predominantly injury-resistant endurance muscle-fiber types. "Some of the best players in the world aren't lightning men," says Galanis. "They're not great athletes. They're actually slow runners."

Even the ones who are great athletes don't win on athleticism. Messi, a four-time winner of the award for the world's best soccer player, has phenomenal acceleration that allows him to zig-zag through crowds of defenders. It makes for terrific highlight reels but masks the fact that the great Argentine does less running than just about anyone else who plays his position at his level. In fact, biometric data from the 2014 World Cup showed Messi ran less per minute of play time than any other player in the tournament except goalkeepers. Trying to understand how it was possible that the seemingly laziest player on the field could consistently dominate matches, the Irish soccer analyst Ken Early speculated that Messi's uncanny ability to read the flow of play allowed him to save steps by anticipating where the ball was likely to end up. "While the others are running just to keep up, Messi only has to walk to stay one step ahead," Early wrote. This reading calls to mind hockey legend Wayne Gretzky's oft-quoted remark, "I skate to where the puck's going to be, not to where it has been."

Galanis still plays regularly, and insists he's better as a soft-bellied 40-something than he ever was in his 20s. "Back then, I'd run around all over the place like an idiot. Now, I'm out of shape or whatever, but I can just walk somewhere and get the ball, have a coffee and pass it. I'm all by myself because I know where the space is. You with me?"

Among Galanis's favorite players to watch is Messi's longtime teammate, the Spanish midfielder Xavi Hernández, who played for Barcelona FC until age 35. As tiny as Messi and not nearly as fast, Xavi (who, like many soccer players, goes by his first name) was regarded by many soccer aficionados as the sport's best passer in his time at Barca. "His whole gig was he thought two plays ahead," Galanis says. "When he passed the ball, he wouldn't just pass it to the next bloke. He'd pass it to that guy in a place where that guy had no choice but to control it on the go . . ." Getting excited, Galanis grabs the mini-ball and acts out receiving a pass on a dead run and having no choice but to play it through to a waiting striker in front of the goal. "It's the quick decisions that make these older players better. It's not their body, it's 'cause they're playing with their mind."

That, in a nutshell, was what Galanis envisioned for Carli Lloyd. It wasn't that he disagreed with Pia Sundhage's assessment that Lloyd had lost a bit of speed, but he believed she had arrived at the place in her development where that kind of speed shouldn't be a factor. In fact, he thought she would be a better player if she learned to rely less on her legs and more on her brain. And that's what they started working on in 2012 with the onset of Phase 3. Their focus was on making plays in what's known as the "attacking third," the 40 meters or so in front of the opponent's goal.

When soccer commentators describe a player's style as "direct," they usually mean it as praise. A direct player is one who drives right at the goal and takes on defenders one-on-one, forcing them to commit themselves or turn and give ground. But Lloyd, he saw, was too direct. The same hunger for the ball that made her great resulted in a lack of tactical nuance. Clamoring for a pass, she would check toward her teammate with the ball, dragging a defender with her and compressing passing channels. Or she would loiter in the penalty area awaiting a cross, giving defenders plenty of time to mark her before the ball arrived.

Hooking his laptop up to an overhead projector, Galanis shows me a series of diagrams in pairs: first, how Lloyd had been playing in a given situation, and second, how he was now coaching her to play it. Some of the corrections are complicated enough that he has trouble explaining them to me, but a couple are quite simple. For instance, rather than sprint directly into the penalty area when a winger advanced the ball down the sideline, he was pushing her to round off the path of her run, looping around behind the center back, in order to arrive just as the ball did and from a less expected angle. It's straightforward enough to diagram, but it goes against all the instincts of an aggressive, direct player like Lloyd. There's a reason Galanis waited seven years to bring it up. "When Carli was a young soccer player getting onto the scene, I couldn't sit there and do these types of descriptions with her, because tactically all she needed to do at that stage was get used to the speed of the game," he says. "The whole idea was to get up to this level, where she is adapted to the quickness of the game and it isn't a factor anymore."

Just how much it isn't a factor became obvious to everyone at the 2015 World Cup. In the final match, again against Japan, Lloyd erupted for three goals — all in the first 16 minutes of play. No one in the history of the World Cup, man or woman, had ever scored so much so fast, and certainly not in such a big game. In fact, only one player, England's Geoff Hurst, had ever scored a hat trick in a final before. But it was the nature of the third goal that elevated Lloyd's performance from record to legend. Receiving a pass near midfield, Lloyd turned with it and took a touch to put some space between her and two nearby defenders. Raising her eyes, she saw Japanese goalkeeper Ayumi Kaihori was well off her line at the outer edge of the penalty area. In a flash, Lloyd struck, sending an arcing shot from perhaps a stride past the centerline. Kaihori, backpedaling, managed to get her fingertips on it but couldn't deflect the shot. Kaihori shouldn't have been so far out of position, of course, but surely she was thinking she would have time to get back before an American attacker could run all the way from midfield into shooting range. She forgot the thing Galanis had finally succeeded in instilling in his pupil: nobody is as fast as the ball. Quick thought beats quick feet.

All that being true, I ask Galanis, where does it stop for a player like Lloyd? Our conversation is taking place a few months after Lloyd's ICBM of a goal and a few weeks after she was named FIFA's women's player of the year for 2015 (an honor she will repeat again the following year). That pretty conclusively marked the apotheosis of Phase 3. Yet if goals and awards were any indication, Lloyd only seems to be getting better. Surely she had reached her peak?

"Your peak ends where you want it to end. There is no peak. There is no peak," Galanis says, repeating that last remark slowly for emphasis. "There is no peak and there is no talent. We make talent. No one is born hitting three-pointers on a basketball court, or bending balls in the top corner. Talent is made."

In fact, he tells me, a sly look coming across his face, Phase 3 was such a success, he and Lloyd recently decided to add to the plan. Would I like to know the goal for Phase 4?

Of course.

"It's to become the best player ever, mate."

Would you doubt her now?

There's a joke Mackie Shilstone likes to tell. You remember Mackie; he's the "career extender" who helped Peyton Manning and Serena Williams win championships in their 30s.

"The old bull and the young bull are sitting on one side of a fence on the top of a hill, and there are these beautiful cows down there at the bottom of the hill, basically pretty women," he says. "The young bull says, 'Let's jump over this fence and run down and go get some of that.' The old bull says, 'How about we just open the gate and walk down and get it?'"

Bodies slow down as they age. Sports science and medicine have gotten better at keeping those bodies healthy, and at restoring them to health when they break down, but they haven't been able to do much about this fundamental fact. Sprinters' race times still drift upward beginning around age 28 at about the same rate they did 50 years ago.

But for athletes in many sports, top-end speed is a little like a policeman's service weapon: you don't want to lose it in case you ever need it, but the better you get at your job, the less likely you are to have to use it. Younger athletes don't run faster because they can; they do it because they haven't learned how not to have to. They haven't figured out how to open the gate and walk down. Although he officially ran a respectable 4.8-second 40-yard dash at the NFL Combine, Manning was one of the less mobile quarterbacks in the NFL even as a youngster. By the time he won his second Super Bowl with the Broncos, he could barely scramble or throw with any velocity, but no one was better at reading the defenders' movements before the snap and anticipating the holes in a coverage. "After fifteen or seventeen years in a sport, don't you think you know what that defense is going to do?" says Shilstone.

At the height of his dominance, as he racked up nine Grand Slam singles titles in five years, Rafael Nadal demoralized opponents with his peerless court coverage, retrieving shots that would have been clear winners against almost anyone else. Nadal's slide from the top of the tennis rankings coincided with a series of knee and ankle injuries in his late 20s. But diminished speed had nothing to do with Nadal's falloff, in the diagnosis of his coach, Francis Roig. If anything, it was the opposite: Nadal was playing like he had less time in points than he really did, rushing to the ball and swinging without setting his feet. Like a pre–Phase 3 Carli Lloyd, he needed to slow down. "In tennis, if you're too fast, it's bad," Roig told the *Wall Street Journal*. "If you're too slow, it's bad. You have to be on time."

"Slow is smooth and smooth is fast." This saying is hammered into the heads of every American Special Forces operator. I first heard it from Eric Potterat, the former head psychologist for the Navy SEALs. When SEALs practice shooting, he told me, they rehearse slowing their movements down so it almost looks like they're moving in slow motion, or underwater. It makes sense when you think about the logistics of a commando mission. Marksmanship matters. You only have the ammunition you can carry with you, and you don't know for sure when you'll be coming home. Wasted shots not only deplete your ammo, they also signal your position to the enemy. And, of course, the more care you take setting up your shot, the more likely you'll hit your target. Better to shoot second but kill first.

"Time is not on his side." I've seen this phrase used in reference to countless older athletes, from Tiger Woods to Usain Bolt to R.A. Dickey to Ben Roethlisberger. I understand what it means, of course, but I disagree. Time is very much on the side of older athletes. It's their best friend. So much of late-career performance comes down to the way expert performers experience time, control it and manipulate it. How many times have you heard an announcer say a veteran quarterback is "a master of using the clock"? The flow of time becomes a weapon: the more the game slows down for you, the more you can make it speed up for others.

There's no clock in tennis, but if there were, no one would use it more effectively than Ivo Karlovic, the Croatian player who holds the all-time record for serving aces, with more than 11,700. When Karlovic ended 2016 as the number 20 ranked man in the world, he became, at 37, the oldest man in nearly 40 years to crack the top 20. There's nothing subtle about Karlovic's game. It's all geometry. At 6 foot 11, he is one of the tallest professional tennis players ever. His height allows him to hit the ball down into parts of the service box shorter players can't reach, at least not without trading speed for heavy topspin. That opens up wide angles of the court unavailable to those who lack a 7-foot wingspan. "When Karlovic serves, the box technically becomes twice as large. The net becomes a foot lower," Andre Agassi wrote about playing him way back in 2005. "It's like he's serving from the freaking blimp." The drawbacks of Karlovic's stature, however, are equally obvious. Extremely tall players tend to lack agility and have trouble digging out low shots, and Karlovic is no exception. His size also means extra stress on joints and tendons. There's a reason more basketball centers don't switch to tennis.

In 2012, when Karlovic was 33, he and his coach, Petar Popovic, sat down to talk about how to get the most out of the rest of his career. The key, they agreed, was to limit his running. "We decided together that he needed to stop completely to play long points and to make a game one, two shots max," Popovic told me. As a human ace machine, Karlovic was already playing a lot of one-shot points, but not enough. Popovic urged the big man to hit harder on his second serve, which he did, increasing the average velocity from 165 kilometers per hour to 190. They also agreed it was worth going for it in his return games, taking more big swings in hopes of ending the point faster, whether with an outright winner or a mishit. Not only has this formula allowed Karlovic to stay generally healthy, his win percentage on second serves has gone up 10 percent, says his coach, and he entered 2017 only a few spots below his career-high ranking of number 14.

Karlovic's success didn't go unnoticed. In the summer of 2015, at the Cincinnati Open, Roger Federer started ambushing opponents with a new time-bending move of his own. On his service returns, every so often, without warning, he would take a couple of quick steps up inside the baseline and hit a half volley, short-hopping the ball and chipping it back while the server was still recovering his balance. It was a tactic that wouldn't have been out of place in the wooden-racquet era, but in the big-serving modern game, It was completely unexpected — a sort of kamikaze attack, except that Federer's peerless touch and ability to anticipate his opponent's serve placement allowed him to pull it off more often than not. Commentators dubbed it the SABR, for sneak attack by Roger. Federer successfully used it in the final at Cincinnati, where he scored a rare victory over Novak Djokovic. In interviews at the U.S. Open a couple of weeks later, he confided that he had come up with the move as a way of, like Karlovic, shortening points and games to limit his fatigue and soreness over the course of the tournament. It also had the effect of discomfiting servers, and their coaches. Djokovic's coach, Boris Becker, objected strenuously, calling it "disrespectful." The SABR worked on all levels, but not quite well enough. In the final against Djokovic, Federer ran out of steam just as he'd feared in the longer best-of-five-sets format.

A few weeks later, baseball fans were treated to their own version of the SABR in the World Series courtesy of New York Mets pitcher LaTroy Hawkins. A 42-year-old reliever, Hawkins had pioneered something he called the quick pitch: from the stretch, he brought his hands to his waist as though he was setting himself, but instead of pausing, he went right into an accelerated pitching motion. Unlike the SABR, the quick pitch wasn't about saving energy. It was designed to take advantage of batters who initiated their swings with a high step of the front foot, timed to the pitcher's motion. Like the SABR, the quick pitch's debut was greeted by accusations of poor sportsmanship and questions about legality. (It's perfectly within the rules as long as there are no runners on base.) It was so effective, Hawkins's younger colleagues in the Mets bullpen demanded he show them how to pull it off. In Game 1 of the championship series against the Kansas City Royals, one of them, 26-year-old Jeurys Familia, attempted to quick-pitch Royals left fielder Alex Gordon. But Gordon wasn't an especially high stepper. Moreover, he was watching keenly as Familia used the ruse to get the batter ahead of him, Salvador Pérez, to ground out. Gordon anticipated what was coming and drove the resulting pitch over the wall for a ninth-inning home run that put the Royals ahead in a series they would go on to win. After the game, Mets manager Terry Collins delivered his verdict: Familia had rushed his delivery. It turns out that even when you're throwing something called a quick pitch, the slow-is-fast rule applies.

If you're an older athlete, complexity is your ally. It's what makes the flow of time feel uncomfortably hurried for those who haven't mastered it and leisurely for those who have. Complexity is what separates experts from novices. Perhaps the clearest example of this is in multisport events like heptathlon and decathlon. The seven events that make up heptathlon — 100-meter hurdles, high jump, shot put, 200-meter dash, long jump, javelin, and 800-meter run — are all to some extent power-based, relying on the kind of quick-twitch muscle fibers that are early casualties of aging. Athletes in all of those individual events tend to achieve their personal career bests in their mid- to late 20s. But peak age for heptathlon as an event is older, around 30. "Generally the more technical a sport, the later one gets in terms of reaching peak," explains Trent Stellingwerff, the Canadian Olympic coach and sports scientist. "If you're doing seven or ten events, it takes a lot of years to master all of those, so peak age tends to be older than for any of the individual events."

You can see it in golf, too. Thanks in no small part to Tiger Woods's then-novel embrace of weight training, golf has gone from being predominantly a sport of technique to one in which the ability to generate power plays an ever more central role. That shift has given rise to a generation of monster-driving youngsters like Rory McIlroy and Dustin Johnson. But pre-Tiger golf returns for a long weekend each summer when the PGA Tour makes its way through the British Isles. Eight of the 10 British Open champions from 2007 to 2016 were over 35. The explanation lies in the wet, windy, short courses that confound players whose games rely on hitting towering 300-yard tee shots. "Links golf" rewards tactical discretion, patience, advance planning, the ability to read conditions and evaluate risk-reward profiles — the things that accrue with years on the tour, not hours in a weight room.

In sports where complexity is on the rise, you would expect to see older competitors having comparatively more success. The advent of so-called advanced analytics has added a layer of strategy to just about every team sport, but it's safe to say no sport has gained new wrinkles as fast as football. The NFL's relatively small number of games, combined with the ever-growing sums of money involved, has turned it into a game-planning arms race. Playbooks that used to resemble pamphlets have ballooned to the size of physics textbooks; now most teams prefer tablet computers for their ever more elaborate schemes. Until 1978, when the league changed its rules about blocking to encourage more passing, the most common form of offense was a simple power running game. Now offenses routinely send four or five receivers out to run patterns, with each receiver making multiple route adjustments based on defenders' reactions. The chief limitation on game-planning complexity is the quarterback, who has to choose from among a mind-boggling number of places he could throw the ball in the three or so seconds he has before getting knocked over.

Oh yeah, and while he's trying to figure out which of his four or five receivers is going to be where, when, he's also trying to avoid pass rushers whose own schemes have grown correspondingly sophisticated. "Nine years ago, if you had five-man protection and they brought five people, there wasn't enough design on defense for them to still get you," Atlanta Falcons quarterback Matt Ryan told ESPN, describing the evolution of his position. "Now defenses are dropping out tackles and ends, bringing certain linebackers on certain sides, all this extra design to make the numbers not right from a quarterback's perspective . . . That's what separates quarterbacks now, the ability to process all that information in a millisecond." What's striking is the degree to which the ones who do it the fastest are also the ones who've been doing it the longest. When Ryan made his first Super Bowl in 2017, at age 31, he was the youngest quarterback of the four to reach the conference-championship round of the playoffs. The others were Aaron Rodgers, 33, Ben Roethlisberger, 34, and Tom Brady, 39. That comes out to an average age of 34.25. The previous year's playoffs, featuring Manning and 36-year-old Carson Palmer, had an even higher mean quarterback age at the conference stage despite the presence of 26-year-old Cam Newton. That's not a fluke but a trend: of the 17 times a quarterback over age 34 has finished the season with a passer rating over 100 in the history of the NFL, 11 have occurred since 2010. It also represents a significant break with football history. Like most quarterbacks of their eras, Joe Montana, John Elway, Dan Marino, and Steve Young all saw rapid performance decreases after age 33. Clearly, something has happened that has tilted the field toward more experienced signal-callers. Recent rule changes that protect quarterbacks from excessively violent hits are part of it, but the bigger part is the ever more cerebral nature of the job.

"You can't surprise me on defense," Brady said in an interview shortly after winning his NFL-record fifth Super Bowl at 39. "I've seen it all. I've processed 261 games, I've played them all. It's an incredibly hard sport, but because the processes are right and are in place, for anyone with experience in their job, it's not as hard as it used to be. There was a time when quarterbacking was really hard for me because you didn't know what to do. Now that I really know what to do, I don't want to stop now."

As athletes accumulate experience within their sports, they become more efficient processors of information. Put more briefly, the longer you play, the faster you think.

But there's intriguing evidence it works in the other direction as well. That is, the faster you think, the longer you'll be able to play — because you'll be able to avoid the kinds of injuries that cut athletes' careers short. That's what a kinesiologist named Charles Swanik has spent the last few years proving.

Swanik, who goes by the nickname Buz, studies the neurological basis of injury-proneness. Elsewhere in this book, we've looked at some of the underlying factors that put athletes at risk for getting hurt: the accumulation of fatigue; range-of-motion limitations and inefficient movement patterns; genetic variations that influence the strength of tissues like bone and collagen; deficiencies in nutrients like vitamin D and magnesium.

Swanik thinks that list is missing a big one: how well your brain models the world around you, and how quickly it responds when that model changes. He began to suspect that was a bigger factor than anyone realized while doing research for his doctoral dissertation, which focused on the effects of injury on proprioception, the body's ability to detect its own position in space. When a muscle or ligament tears, so do the nerve fibers embedded in that tissue. But nerve cells regrow more slowly than other tissues, often requiring a year or more before they recover full function. A badly sprained ankle or knee may be mechanically sound after a few months of rehabilitation, but the feedback it sends to the brain about its position and the forces being placed on it is incomplete. That's one reason the biggest injury risk factor of all is previous injury.

If a communication breakdown between brain and limbs is responsible for many reinjuries, Swanik reasoned, might it play a similar role in new injuries? Tissue damage, he knew, happens in the blink of an eye: from the time loading forces are applied to when a sprain or strain occurs is only 70 milliseconds. That's faster than a human reflex, which requires about 80 milliseconds. Every time we encounter a sudden force and *don't* get injured, then, it's not because our lightning-quick reactions are saving us, but because our brains were able to anticipate the force and coordinate an appropriate muscle-activation strategy.

"Before an injury happens, the brain has already planned out and modeled that person's physical surroundings, the local environment," Swanik told me. "But if something changes in the middle there, you essentially are executing some kind of strategy that doesn't match your environment." In his lab at the University of Delaware, Swanik has conducted a series of experiments to show what happens when people are caught by surprise in the middle of an action. He has asked blindfolded subjects to jump onto a platform without knowing its height or land on an unstable surface. Electrodes attached to their muscles captured the results. In a normal jump landing, the muscles fire in a smooth cascade that transfers force harmlessly up the kinetic chain from ankles to knees to hips. An "unanticipated landing," however, typically results in a startle response, in which muscles tense and release in a random, uncoordinated way; think of how your body simultaneously buckles and jerks when you step into a pothole you didn't see. "When an injury happens it's always so surprising because literally the brain did not anticipate what was about to happen," Swanik explains.

Swanik knew that athletes who suffer concussions are subsequently at higher risk for all sorts of musculoskeletal injuries, not just more concussions. What if, he wondered, it's not the concussion itself that's to blame but impaired cognitive functioning that makes the brain too slow to react when something unexpected happens — the ball takes a deflection, the player you're guarding throws a sick shoulder fake, that sort of thing — resulting in an awkward startle response? If that were the case, he thought, nonconcussed athletes who are innately slower processors should also exhibit a higher injury risk.

To test this hypothesis, Swanik and his research team administered tests of neurocognitive function to nearly 1,800 college athletes. The athletes, who included football, soccer, lacrosse, and field hockey players, took computer tests that measured their memory, reaction times, and processing speed. Over the ensuing season, 80 of the 1,800 athletes sustained noncontact ACL injuries. Compared with the control group of demographically identical noninjured athletes, the ones who tore their ACLs, it turned out, had on average performed significantly worse on the cognitive screen, demonstrating slower processing, longer reaction times, and worse memory.

"That was kind of proof for me that you can have the biggest muscles and whatever, but if you neurologically have a breakdown somewhere in this system, you're going to injure yourself," Swanik told me. "All this time we've been measuring biomechanics, which is really an outcome. What we should be looking at more is the processing, the underneath neurological strategy that resulted in the biomechanics we saw."

All of the above suggests that, all else being equal, the kind of expert veteran athletes who use high-level chunking to process game situations efficiently should be better at modeling their environments and avoiding injuries of uncoordination than less-experienced younger players.

That's exactly what Swanik suspects. "If you accumulate that kind of sensory experience over time, that gives you a better prediction of what you need to do next," he says. Often, he notes, the goal of strategies in sports is explicitly to negate the advantage of experience by taking away the ability to process data as chunks. When football defenses disguise their coverage and blitz schemes or present unscouted looks, they're trying to slow the quarterback down enough that a pass rusher has a chance to surprise him and force an uncoordinated response like a fumble or a bad throw — or an injury.

All else isn't equal, of course. Older athletes don't just have bigger mental databases to draw from. They also have the physical baggage that goes along with it — fewer quick-twitch motor units, slower-recovering muscles, more general wear and tear. It's also worth remembering that age and expertise are two different things. There's little or no carryover of expertise between sports, so taking up a new one at 40 carries that much more injury risk than doing it at 20.

Still, if you've ever wondered how athletes like Tom Brady or Carli Lloyd or Jaromir Jagr are so successful avoiding injuries after 20 years in a high-intensity sport, it appears at least part of the answer is just that: because they've been doing it for 20 years.

9

CAREFUL WHAT YOU SWALLOW

*The Fads and the Facts of Career-
Extending Nutrition*

On a muggy May morning, the alarm on my iPhone goes off two hours earlier than usual and I will myself to sit up in my hotel bed, feeling like wet garbage. I'm three days into the worst flu of my life. On top of that, I've spent 12 of the last 48 hours on an airplane, flying from New York to San Francisco, then to Los Angeles, and finally, last night, into New Orleans. I would kill to chug some NyQuil and pull the covers over my head. But when the most decorated African athlete in Olympic history invites you to join her for a workout, you can't just bail.

Twenty minutes later, I walk into the fitness center at the Intercontinental Hotel and say hello to Kirsty Coventry. Dressed in a pink tank top and black tights, the 32-year-old Zimbabwean swimmer has already begun her warm-up: 14 minutes on the exercise bike, alternating every minute between pedaling hard and taking it easy. With her flaxen hair, blue eyes, and tall, lean build, Coventry could have played an elf in *Lord of the Rings*. Representing her continent at the 2004 and 2008 Summer Games, she brought home seven medals overall — four of them in the backstroke, and three of them gold. No Olympian from another African nation has as many, and no woman swimmer from any nation has more. In Zimbabwe, where she is a national celebrity, Coventry's nickname is Didi Mukuru — the Great Swimmer.

Slathering my hands in Purell — with Rio three months away, giving her my flu right now could be disastrous, I know, although Coventry seems unconcerned — I hop on the bike next to her. Fortunately for me, Coventry's schedule yesterday included a rigorous pool session back home in Atlanta, where she lives during training, so this morning is just a light maintenance workout. Light for her, anyway. After finishing up on the bikes, we do a circuit of floor exercises, mostly core stuff, and then another that involves a complicated one-handed clean-and-jerk. She uses a 35-pound dumbbell; eyeing her domed shoulder muscles, I select a 15. Coventry proves to be a delightful, encouraging workout partner and getting my heart rate up leaves me feeling better than I have in days.

We head downstairs for breakfast. Knowing how particular elite athletes can be in their diets, especially the ones over 30, I'm half-expecting Coventry to produce a Tupperware container out of her gym bag, or canisters of various powders, or perhaps to summon a chef to take her complicated special order, but she joins me at the buffet, and when we meet at the table, her plate looks pretty much like mine: omelet, potatoes, fruit. In between bites, she tells me about the preparations for her third Olympiad.

It was supposed to be her fourth, but during her training for the London games in 2012, she suffered a double misfortune: a dislocated kneecap followed by pneumonia. It was a bitter disappointment and she needed time away from the sport to get over it. She ended up taking two years off, moving home to plan a wedding with her fiancé, Tyrone Seward, who observed the traditional South African tradition of *lobola,* paying Coventry's father a cow and two chickens for his daughter's hand.

Coventry enjoyed her time off, taking boxing classes and going hiking with Seward, getting in the pool only occasionally. But the idea of ending her competitive career with a misfire didn't sit well with her. "I love the sport, and I love my career, and I wanted to walk away being like, 'I'm proud and I'm OK now to walk away,'" she says. "Instead of being like, 'This sucked.'"

She had never officially retired but was nevertheless anxious before calling her coach, Kim Brackin, to sound her out about a comeback. "I probably took two or three weeks to pluck up the courage to call her, because she and I have a very honest relationship. And I knew that if she was hesitant, then I'd be like, 'Oh, she doesn't think I can do it.'" But Brackin's response was immediate: Coventry still had it in her to be world-class.

It didn't feel like it when she moved back to Atlanta and resumed training. "The first six to eight months were miserable," she recalls. "I would go home and be like, 'Why the hell am I doing this?'" But any worry she had that she might be too old was dispelled when she looked around and saw what her contemporaries were doing. With occasional exceptions — like American Dara Torres, who won three silver medals in Beijing at age 41 — swimming has long been one of those sports where careers tend to peak in one's early 20s. But the reasons for that are at least as much financial and psychological as physiological. It's hard for all but a handful of the top swimmers to earn a comfortable living, and just as hard to spend countless hours of your life staring down at the black stripe on the bottom of a pool. As Rio drew nearer, Coventry was heartened to see that many of her peers were defying the expectation that they wind down their competitive careers by 30. Michael Phelps would be 31 when the games began. Ryan Lochte would be 32, and Natalie Coughlin 33 (although Coughlin, somewhat shockingly, would end up failing to qualify for Rio). "There's a considerable big bunch of us that are twenty-nine and older and doing well," Coventry says. "People are starting to say, 'Oh, we can still do this — we just need to change the way we train.'" For her, that mostly means paying more attention to her fatigue level, varying the intensity of her workouts and making sure to give herself a day off when in doubt. "At twenty-two, I would be able to go into the pool every single day, have great practices, come back the next day and be fully recovered. And now that recovery process is slower," she says.

That's not to say she's taking it easy now. Most days, Coventry spends a couple of hours in the pool in the morning and hits the weight room in the afternoon. On her heaviest day, Thursday, she does no fewer than four workouts: a pool session, a hot yoga class for active muscle recovery, weight training, and then another swim practice.

Burning so much fuel, Coventry doesn't worry about how much she's taking in. In fact, she doesn't know how many calories she ingests in a day, a fact that surprises most people when she tells them. Instead, she focuses on when and what kinds of food she's eating. To keep her energy level high and even, she eats small meals and snacks throughout the day, never going more than two or three hours without grazing. Before and after practices, she'll have something with protein, whether a shake, an energy bar, or a handful of nuts. She loves red meat but saves it for dinner, finding it easier to swim in the afternoon if she has chicken or fish for lunch. She avoids fried foods. She tries to eat organic but isn't rigid about it. She's certainly not one of those athletes who thinks sugar is poison. "I have a crazy sweet tooth," she says. "I love chocolate, I love candy, I love ice cream. I'm not ever going to give it up. But I have it little bits at a time."

I'm just about to write off Coventry as a boringly commonsense eater when she casually mentions that the only thing she won't eat is anything genetically modified. No GM corn or soy protein, nothing cooked in GMO canola oil. It started, she says, in 2012, the year she missed the Olympics. Coventry and her coach were in the weight room, running side by side on treadmills. "We're both really competitive. She was like, 'Don't let me catch you' and I was like, 'Oh, heck no, you're not catching me.' And so I started sprinting." Like most top swimmers, Coventry has exceptionally limber joints. In the water, that allows her to make longer, more efficient strokes. On land, though — well, the next thing she knew, her patella had come unmoored from the middle of her knee and drifted to one side of her leg.

For the next few weeks, Coventry was stuck doing rehab several hours a day, stroking only with her arms while a buoy held up her immobilized legs in the water. Maybe she was pushing herself too hard in compensation, or perhaps the stress of the injury had compromised her immune system, but two months after her knee injury she was back in the hospital, this time with pneumonia. The sight of a strappingly healthy 28-year-old checking herself into the hospital in the middle of the summer with an old-person's ailment was so strange, she recalls, the nurses were laughing about it and then apologizing. Coventry was ordered to go on bed rest for 10 days, and that was pretty much it for her hopes of swimming in London.

Devastated, she went looking for answers. "The one thing we could control at that point was the food," she says. From her sickbed, Coventry read up on foods that were supposed to be healing and consulted friends on what had worked for them. The idea to stop eating foods made from organisms with artificially altered DNA originated with her friend Therese Alshammar, a Swedish swimmer who, at 38, was training for her sixth Olympic appearance. Foods that don't occur in nature, she said, are harder for the body to digest. Eating them, therefore, temporarily saps one's energy in the same way as eating a big steak at lunch. Coventry wondered: Could GM foods in her diet have contributed to the immune system lapse that brought on her pneumonia?

Outside the U.S., suspicion of GMOs in food is common; in Europe, genetically modified crops are heavily regulated and foods containing them require special labels. Coventry isn't one of those people who thinks GMOs are dangerous or some kind of corporate conspiracy, but Alshammar's advice made sense to her and tallied with the way she thought about food generally. If she was already choosing quinoa over bread for her carbohydrates because the former is less processed and closer to something our hunter-gatherer ancestors might have consumed, why eat grains or produce containing DNA from a laboratory? "For me, just anything modified is not something that our bodies are used to eating. It's all about trying to be as efficient as you can and allow your body to get in as much goodness as it can without having to overwork," she says. "And it's working."

In fact, while Coventry's post-2012 nutrition strategy has undoubtedly worked for her — in Rio, she made it to the finals in the 200-meter backstroke, although she didn't add to her medal collection — there's no reason to think avoiding GMOs has anything to do with it. Hundreds of scientific studies over the years have established that not only are common GM foods safe for humans and animals, they're nutritionally indistinguishable from non-GMO versions. In 2016 the U.S. National Academies of Sciences, Engineering, and Medicine attempted to put the debate to rest by releasing a 388-page report overseen by a committee of more than 50 scientists and food-science experts. It categorically declared the most common allegations aimed at GM crops — that their proliferation has been responsible for increases in cancer, allergies, obesity, autism, and gastrointestinal disorders — to be unfounded. An independent and highly regarded watchdog, the Center for Science in the Public Interest, praised the report's thoroughness. To be sure, if you go looking on the internet for studies suggesting GMOs are bad for you, you can find some. And, to my knowledge, no study has ever looked specifically at the question of whether modified foods make athletes feel logy. But it's a firm consensus of the scientific establishment that GMO crops are safe and just as nutritious as their nonmodified equivalents.

In theory, it ought to be surprising to hear a smart and down-to-earth world-class athlete attesting to a theory that's been fairly conclusively debunked. In reality, nothing could be less remarkable. Welcome to the weird world of sports nutrition. In no other area of sports science do quasi-scientific fads, fashions, and even outright frauds flourish quite so abundantly. And older athletes, always looking for another trick to turn back the clock, are the most receptive audience for all of these things.

A certain longtime NFL starting quarterback not famous for his healthy habits has been known to put shots of his wife's breast milk in his smoothies, believing whatever is in there that makes babies grow and heal so fast must be good for him, too. Amar'e Stoudemire, the longtime NBA player, told me his kosher diet was a major factor in his comeback from the ill-advised microfracture knee surgery that almost ended his career. (Conveniently, Stoudemire now plays in Israel.) In his final season, Kobe Bryant never played a road game without making sure the chef at the Lakers' hotel knew how to prepare bone broth to his specifications; the viscous traditional soup, he said, helped keep his cartilage from wearing down. Novak Djokovic has eaten a strict gluten-free diet ever since a traditional Chinese medicine practitioner (who also specializes in magnet therapy and the "negative influence" of "geopathic radiation") convinced him he had an allergy by having him hold a piece of bread against his stomach to see if it made his arm weak. Dwight Freeney, the great NFL pass rusher, once prepared for the playoffs by eating nothing but beef and pinto beans for days on the advice of a nutritionist who claimed those were the foods that were most compatible with his immune system that month. Dozens of NHL players have sought the advice of Gary Roberts, a

Toronto performance coach who claims he can maximize players'
longevity with the help of a diet heavy on goat's milk, mung beans,
and hemp.

Surely the athlete most famous for touting the career-length-
ening benefits of a weird diet is Tom Brady. With five Super Bowl
rings, the New England Patriots quarterback has a strong case as
the greatest football player ever. Brady was a 24-year-old rookie
when he won his first championship; he won his fifth at 39, and
might have secured his third MVP award for the season if not for
the four-game suspension he served for a conspiracy to underin-
flate footballs. (If you don't already know, it's a long story.) At a
press conference after Super Bowl LI, Brady credited his longevity
to the work he does with a holistic guru named Alex Guerrero, who,
for the past decade, has managed all aspects of his health. "When I
was twenty-five, I was hurting all the time, and I could never have
imagined playing this long," Brady said. Now, "I'm thirty-nine, and
I never hurt. My arm never hurts, and my body never hurts. I know
how to take care of it."

Brady and Guerrero are now partners in a holistic wellness clinic
called TB12. In 2016 they published *The TB12 Nutrition Manual*,
a collection of 89 "seasonally inspired" recipes modeled on Brady's
own Guerrero-designed eating plan. Dishes include gnocchi made
from sweet potatoes (Brady doesn't eat regular potatoes or other
"nightshades") and avocado ice cream (Brady doesn't eat sugar or
almost anything sweet). With a $200 price tag and laser-etched
wood covers, it was pitched not as a cookbook per se but as a "living document"; buyers were promised the purchase would include
future batches of recipes. It sold out almost instantly.

Other ingredients you won't find in *The TB12 Nutrition Manual*, should you get your hands on a copy: wheat flour, dairy, straw-
berries, mushrooms, iodized salt, or MSG. Brady doesn't drink coffee and rarely touches alcohol. Like Kirsty Coventry, he doesn't eat
genetically modified foods. For protein, he relies mostly on grass-

fed steak, wild-caught salmon, and duck. When Brady traveled in China after his fifth Super Bowl victory, he packed all his own meals rather than risk deviating from his program. At the core of it all is a balance of 80 percent "alkaline" foods to 20 percent "acidic" foods. Guerrero subscribes to a belief common among alternative healers that acidic foods promote inflammation by lowering the pH balance of the blood. "The athletes that we deal with as our clients notice a big difference when they cheat on our diet," he assured me.

That's entirely plausible. The question is whether the mechanism responsible for the difference they feel is what Guerrero and Brady believe. Proponents of alkaline eating for athletes can point to a couple of biological facts. The chemical reaction that takes place in muscle cells during intense exercise produces free hydrogen ions, making the intra- and intercellular environments more acidic. The buildup of this acid during anaerobic metabolism is what causes short-term muscle fatigue; temporarily raising the pH of blood by consuming large amounts of an antacid like sodium bicarbonate has been shown to delay the onset of muscle fatigue. Meanwhile, chronic acidosis of the sort that occurs in diabetics with renal disease causes accelerated bone loss similar to that seen both in the elderly and in high-volume endurance athletes. Dietary supplementation with alkaline calcium salts has been shown to moderate this bone resorption.

It's a long way from there, however, to the claim that acid-promoting foods are a major culprit in ordinary inflammation, soreness, and injury, as Brady and Guerrero believe. A 2012 literature review in the *Journal of Environmental and Public Health* examined various claims associated with alkaline eating and found qualified support for the notion that such a diet could ameliorate common problems of aging including muscle wasting, bone thinning, and even back pain. In particular, the author noted, magnesium, an alkaline mineral, is a critical component in many enzymatic processes. Magnesium is required for the metabolism of vitamin D, whose absence increases risk of bone and muscle injuries. There's also evidence that neutralizing excess acid helps remedy very low levels of human growth hormone.

But the review also noted that the human body "has an amazing ability to maintain a steady pH in the blood." Whether you're eating like Tom Brady or like Morgan Spurlock in *Super Size Me*, your blood pH is almost certainly very close to a slightly basic 7.4. The effects of alkaline supplementation are pronounced in people with renal disease and postmenopausal women, but for the rest of us they're not so clear. Most nutritionists who have looked at the Brady diet have concluded that the lion's share of its benefits stem from the fact that the most common alkaline-rich foods — green, leafy vegetables, root vegetables, cabbage, beans, avocado, squash — are foods we should be eating anyway because they're rich in all sorts of nutrients. Meanwhile, the diet's more exotic recommendations, like its prohibition of coffee, tomatoes, some berries, and all dairy, including yogurt, are not only lacking in scientific support; they run counter to recommendations of the vast majority of sports nutritionists and doctors.

In other words, even with its unnecessary exclusions, the Brady diet is a healthy way of eating chiefly because of what it's not: a typical American diet loaded with sugar, saturated fats, preservatives, and way too many refined carbohydrates. Anyone who gets 80 percent of his calories from minimally processed vegetables and complex starches, as Brady does, is doing pretty well in the nutrition department, whether or not the other 20 percent involves cheese or tomatoes. When you look closely enough at any trendy diet, from the various flavors of "Paleo" programs that seek to reproduce the eating habits of prehistoric human ancestors, to "color diets" that require eating foods of different hues depending on the day of the week, that's the common thread. Most people who switch to them see a benefit because most people, even athletes, have not-so-great eating habits to begin with. It's more about the diet they're quitting than the one that replaces it.

Even if the 80-percent-alkaline thing were some kind of magic formula for suppressing generalized inflammation, that wouldn't necessarily make it an ideal diet for athletes. Certainly, researchers in the last few years have established links between chronic low-level inflammation and diseases like cancer, diabetes, and Alzheimer's. But even in those diseases, the causality and exact mechanisms are far from well understood. The idea that system-wide inflammation is to blame for achiness, low energy, injury susceptibility, or other common complaints of aging athletes is, scientifically speaking, little more than a hunch, albeit one that's attracting a lot of interest right now.

Set against that is the certainty that the body's inflammatory response is a necessary step in the process through which muscles adapt to training stimuli. Athletes who want to tamp down inflammation can already do so easily enough without the aid of a $200 cookbook. They can take nonsteroidal anti-inflammatory drugs like ibuprofen, or consume large quantities of certain foods or supplements that are high in antioxidants, molecules that disrupt the inflammatory cascade. Tart cherry juice, watermelon juice, and extract from an Indian fruit called kokum have all been shown to diminish postexercise soreness. The problem is that, in so doing, they're also diminishing the magnitude of the body's adaptive response to that exercise, says Asker Jeukendrup, a sports nutritionist and physiologist who has advised many of the world's top teams, including FC Barcelona, the Rabobank cycling team, and the British Olympic Association. Jeukendrup often encounters athletes who think taking "superdoses" of nutrients like vitamins C and E will help them rebound from injuries. He advises against the practice.

"People often think of antioxidants as something healthy, something you need," says Jeukendrup, an avid cyclist who's been keeping training journals since he was 11. "If I've just done a really hard workout and I know my muscles are going to be sore, maybe taking some antioxidants is going to help me tomorrow. But if I take a high dose of that, and then after my workout tomorrow I do the same, and I do it day after day, ten weeks from now I'm going to have an effect much smaller than if I hadn't taken those supplements." That is, you won't get as strong. Jeukendrup suggests reserving NSAIDs, tart cherry juice, and the like for periods when fast recovery is paramount, such as during multiday tournaments.

Although they're susceptible to urban myths and wishful thinking, older athletes do tend to be smarter about what they put in their bodies than young ones, Jeukendrup says. "Hardly ever in young athletes is a diet ideal or optimal. Diet is not at the forefront of their thoughts at all."

In terms of their nutritional needs, though, younger and older athletes are almost identical — with one major exception. "There is one area where we have quite a bit of evidence there may be a difference and that has to do with maintaining muscle mass, with protein synthesis," Jeukendrup says. "With aging, there are some changes in protein metabolism. It's called anabolic resistance. They become a little bit more resistant to increasing their muscle mass." To counteract the effects of anabolic resistance, he says, athletes should increase their protein intake the older they get. One of the components of protein, the amino acid leucine, is particularly important; foods high in leucine include chicken, beef, fish, soy, and — sorry, Tom Brady — cheese.

Most sports nutritionists already agree that athletes of any age need a lot more protein than sedentary adults and often aren't getting it. In fact, the emerging consensus is that very active people need about twice as much, in the neighborhood of 0.6 to 0.8 grams per pound of bodyweight daily. But that doesn't mean you should be doubling the size of the salmon fillet on your plate. According to Jeukendrup, the timing of protein consumption is just as important as the amount. Ingesting more than 25 or so grams of protein at a time merely leads to formation of excess urea, a by-product excreted as urine which plays a role in the formation of kidney stones. It's better to, like Kirsty Coventry, eat smaller amounts of protein every three hours throughout the day. Moreover, there's compelling data showing that protein consumed right before going to sleep is particularly effective at boosting protein synthesis. Maybe it's time to bring back the warm glass of milk at bedtime?

In addition to eating more protein, there's one other nutritional intervention that shows significant promise for older athletes in particular: gelatin. Recent studies have provided evidence that consuming it helps prevent and heal a variety of soft-tissue injuries. The science behind it seems to be absurdly simple in the way science rarely is. Gelatin is made by boiling down the carcasses of cows, pigs, and sheep to liquify the collagen in their ligaments, tendons, cartilage, and skin. Eating that collagen supplies proteins that enable the body to form its own new collagen, which is not only the major component of tendons and ligaments but also of the tissue matrix that binds individual muscle fibers to each other, allowing them to transfer force without tearing. Studies have shown gelatin supplementation during injury rehab results in faster return-to-play after ACL reconstructions and Achilles tendon ruptures.

The leading proponent of gelatin as sports nutrition is Keith Baar, a biologist and physiologist at the University of California at Davis who oversees a lab that focuses on "the molecular determinants of musculoskeletal development." Baar says the degradation of intramuscular collagen that occurs with age is to blame for a phenomenon every athlete over 30 has experienced: ever-greater postworkout muscle soreness from ever-smaller exercise loads. He recommends taking a dose of gelatin an hour before your workout for maximum collagen-building. It doesn't seem to matter what form you ingest, whether it's good old Jell-O, the hydrolized version that blends into smoothies, or capsules.

Alternatively, you can get similar benefits from drinking bone broth, like Kobe Bryant. Yes, the traditional soup craze the Black Mamba brought to the NBA is one sports-nutrition fad with real science behind it. And, bonus, bone broth is delicious. The downside is the preparation time: to make Bryant's bone broth, the Lakers' team chef boiled chicken carcasses for a full eight hours, until the connective tissue had all become "liquid gold." If that sounds like too much commitment, you can do what Baar had athletes on England's Olympic team do in 2012: anytime you have chicken or turkey, swallow as much of the cartilage, gristle, and even bones as you can chew up. Baar calls this the "hyena diet." Have you ever seen a hyena with bad knees?

Most sports nutritionists think diets like Tom Brady's benefit athletes because they promote general healthy eating habits, not because they have any miraculous effect on inflammation or other physiological processes. But it's hard to ignore how many elite athletes report feeling transformed — stronger, less achy, more energetic; in a word, younger — after cutting out one or more specific foods from their diets. The problem is there's no agreement on which foods ought to be avoided, and scientific efforts to pin the blame on any individual culprit invariably come up short.

But what if that's because we've been looking for the wrong thing? What if one athlete's healthy, nutritious meal really is another's energy-sapping inflammation bomb? In theory, there's nothing terribly radical about this idea. About 1 in 25 American adults has a food allergy, and lactose intolerance, the inability to digest a protein found in cow's milk, is even more common, occurring in a majority of people of Asian and African descent. But the idea that there's another category of food "sensitivities" that don't rise to the level of allergy or intolerance but nevertheless have palpable health effects is controversial.

It's getting less controversial, however. For years, it was an article of faith among mainstream doctors and nutrition scientists that cutting out gluten, a protein found in wheat, was only of benefit to people with celiac disease, which afflicts 1 percent or less of the population. But as the gluten-free eating trend that began with the popularity of the low-carb Atkins diet in the early 2000s refused to abate, scientists revisited the question — and concluded that a majority of the nonceliac gluten abstainers could have some sort of previously unrecognized sensitivity. "There is a large placebo effect — but this is over and above that," Benjamin Lebwohl, director of clinical research at Columbia University's Celiac Disease Center, told the *Washington Post* of the new finding. You can understand why it was so hard to isolate; symptoms of the syndrome include fatigue, intestinal discomfort, and mental fogginess — the sorts of things anyone with less-than-ideal sleep or eating habits experiences. (I know I do.) There's still no consensus about how the sensitivity works, or even whether it's gluten that causes it; some researchers believe a type of carbohydrates called FODMAPs, which are also found in dairy, are the more likely suspect.

If it turns out our responses to different diets are far more individualized than we know, as is looking like the case, genetic variation plays a large part. We already know how some of this works. After I sent a vial of my spit to 23andMe for genotype analysis, I got a report back saying my genes suggest I don't have a problem digesting lactose (correct), I'm a slow caffeine metabolizer who doesn't drink a lot of coffee (this was accurate until I became a parent), and the amount of saturated fat in my diet is unlikely to have a large effect on my weight. Plugging my DNA data into Stuart Kim's online genotype-analysis engine showed I'm right in the middle of the curve when it comes to how my body uses vitamin D, vitamin E, magnesium, and iron.

A host of startups are trying to package insights like these for use by athletes. Jeremy Koenig was a national-level sprinter in high school who went on to earn degrees in biochemistry and molecular biology en route to a professorship in something called nutrigenomics at Mount Saint Vincent University. In 2014 he founded Athletigen, which uses genomic data like the report I got from 23andMe to offer diet, training, and injury-risk recommendations to athletes. Koenig thinks genetics will soon solve the mystery of why certain diets work so well for some people and do nothing for others. But that, he says, will only be the start. He envisions using DNA tests to tailor athletes' meal plans to their individual nutritional needs in the way runners like Meb Keflezighi already rehydrate during races with beverages that replace the exact mix of electrolytes they sweat out. We know about the Paleo Diet and South Beach Diet, Koenig says. "What about the Jeff Diet? What does that look like?" (Eric Topol, a prominent genetics researcher and cardiologist at the Scripps Institute, says this scenario is theoretically plausible but won't be realistic until billions of people have had their full genomes sequenced.)

More speculative still are claims that food sensitivities are not only idiosyncratic to each of us but in constant flux. Dwight Freeney, the NFL star who restricted his diet during the playoffs to pinto beans and beef, did so on the advice of a Miami nutritionist named Sari Mellman, who had told him those were the foods that were most efficient for his immune system during that window of time based on an analysis she performed on samples of his blood. When I tried to get in touch with Mellman, I heard back from her son, Leon Mellman, who took over his mother's practice after she died in 2008. Mellman sent me a nondisclosure-agreement form, which he required me to sign before he would speak with me, and, for some reason, a cartoon of an anthropomorphic carrot wearing a top hat.

Despite my agreement in writing not to disclose any trade secrets, Mellman repeatedly evaded my questions about the science behind his mother's program over the course of our nearly two-hour conversation. The answers to my questions, he said, consisted of intellectual property so valuable, large pharmaceutical companies he also declined to name had resorted to corporate espionage, sending disguised agents into NFL facilities to steal materials from Mellman's clients. Mellman was happy enough to name some of those clients — they have included retired wide receiver Cris Carter and running backs Jerome Bettis and Ricky Williams — and tell me in detail about his biography, including the episode that inspired his mother's original research into nutrition and immunology, Leon's near-fatal allergic reaction to apple juice when he was four years old.

As for the program itself, I only got the bare outline: every few months, Mellman clients submit to extensive blood panels designed to tease out how their immune systems react to a wide variety of foods. "If you get drawn I can tell you the five most efficient foods for your immune system," he said. A blood panel, for instance, might reveal that I should be eating lamb but shouldn't be eating beef, even though they're both red meats, based on their differing effects on prostaglandin, a hormone-like fatty molecule that regulates inflammation. "You have to understand how to pair up a food structure with a white blood cell and a white blood cell with an organ system," he said elliptically. "There are over sixty mechanisms within how white blood cells interact with food structure." I didn't bother with a follow-up call.

It's true that immune system function can have an effect on how the body processes nutrients. Hilary Stellingwerff, the Canadian middle-distance runner, was puzzled by a collapse in her performance that occurred in the spring of 2008, just as she was attempting to qualify for that summer's Olympic Games. After a series of perplexingly bad races, Stellingwerff learned in July — after it was too late to do anything about it — that her blood was low in iron. Blood tests traced the condition to a surprising cause: spring pollen allergies. Stellingwerff, who has asthma, knew she had pollen allergies, but she and her husband were both surprised to learn they could affect how efficiently her blood carried oxygen. "Here I am the nutrition and physiology guy and my wife becomes anemic," Trent said.

But the blood testing being peddled to elite athletes — as well as to recreational athletes and dieters with disposable income to spare — goes well beyond nutrient deficiencies, claiming to identify potential performance-sapping food reactions that are too subtle to produce obvious allergy symptoms. Blood panels like the ones Leon Mellman uses screen for the presence of antibodies that supposedly show the immune system is mounting a response to an allergen. The antibody immunoglobulin E, or IgE, is known to mediate allergies to foods like milk, eggs, and nuts. Some popular commercially available blood panels also screen for the presence of immunoglobulin G (IgG). The American Academy of Allergy, Asthma and Immunology (AAAAI) and its European and Canadian counterparts have declared IgG testing not a valid tool for detecting food sensitivities. The presence of IgG in blood serum in response to a food, they have said, only indicates that there has been exposure to that food, not that the body is mounting any sort of immune response. Treating it as evidence of a food sensitivity only results in people eliminating healthful foods from their diets when the goal should be to eat with diversity. Even the presence of IgE should be disregarded, according to a 2010 AAAAI position paper, if there's evidence the food is being tolerated well. In other words, if you're surprised to learn you're allergic to a food you've been eating, it's probably not a real allergy.

Yet many athletes remain convinced blood screening can help them perform better and longer. I heard this view from Brendon Ayanbadejo, who played linebacker for the Baltimore Ravens, Chicago Bears, and Miami Dolphins, retiring after winning a Super

Bowl with the Ravens in 2013. Ayanbadejo was 36 during his final season, and statistically it was his best. That's no coincidence,
he told me. After a torn quadriceps muscle in 2009 cost him most
of a season, he, like Kirsty Coventry, went deep on nutrition and
sports medicine in an effort to keep it from happening again. He
paid out of pocket for platelet-rich plasma (PRP) therapy, which
involves putting blood in a centrifuge to isolate pro-healing and
anti-inflammatory factors and injecting them into the injury site.
He had injections of stem cells, immature cells that can become
any kind of tissue. He experimented with superdosing of vitamins
— the practice that Asker Jeukendrup advises against — and went
on an elimination diet, paring his intake down to five basic foods
and gradually reintroducing things to see what made him feel better or worse. "I paid tens of thousands of dollars a year to stay on
top of my health because the ROI was a million dollars," Ayanbadejo says. "For me, if I have something that's going to take a
month or six weeks to heal, if I need something to heal it in a week,
I'm going to get it." Then, at the recommendation of his teammate
Ray Lewis, he flew to Phoenix for a consultation with Thomas Incledon, a nutrition specialist and strength coach who has helped
a number of famous athletes manage their health late in their careers. Incledon screened Ayanbadejo's blood against 300 different
foods and told him he should avoid blueberries, which, he said,
were responsible for the weight fluctuations Ayanbadejo had been
experiencing since his injury. He told him it might be a transient
sensitivity rather than a permanent one and recommended a follow-up screening in six months. Ayanbadejo did as instructed and
his weight stabilized.

When I went to see Incledon in person, he had a lot of stories
like this to share. Over the course of several hours at his clinic, located in a strip mall in Scottsdale, and two meals at nearby Italian
restaurants, he told me about one star player after another who
came to see him for mysterious pain or health problems that re-

sisted conventional treatment and left whole again after Incledon spotted the nutrient deficiency or hormone imbalance all the doctors had missed.

Even in the world of human-performance science, where it's common to wear a lot of hats, Incledon is a magpie of specialties. According to his website, "Tom's academic background includes an A.S. in Management, B.S. in Exercise Science, B.S. in Nutrition, M.S. in Kinesiology, and Ph.D. in Exercise Physiology. He is a registered dietitian, certified strength and conditioning specialist (CSCS), and plans to pursue a medical or naturopathic doctor license in the future."

Incledon has also billed himself as "The World's Strongest Scientist." A former competitive weightlifter — again, according to his website — "he has been ranked nationally in powerlifting, Olympic lifting, all around lifting, and strongman competitions. His best lifts include: 187 pounds in the one arm snatch (with 7 foot bar), 286 pounds in the power snatch, 352 pounds in the power clean and jerk, 452 in the front squat, and 615 pounds in the dead lift, all while weighing under 200 pounds. Tom has also set national records in strongman competitions, including 19 reps in the 200-pound axle press." An old media kit shows Incledon in a harness pulling a tractor-trailer and deadlifting the back of a pickup truck. His strongman days are behind him, thanks largely to an injury he sustained at a contest when a loose truck tire smashed into his knee from the side. But in a shiny blue athletic shirt and basketball shorts, he still looks as solid and immovable as the concrete-filled oil drums he used to hoist.

At his clinic, Human Performance Specialists, Incledon offers services from acupuncture and body-fat testing to PRP and something called prolotherapy, an alternative medical treatment in which sugar water is injected into joints, tendons, and ligaments to promote tissue regeneration. As with many alternative procedures, the evidence showing prolotherapy works is somewhat patchy and anecdotal, but it doesn't seem to cause harm, at least. There's an infrared sauna and something called the Jade Fuzion 8000, which uses heated jade stones that have "relaxing, healing warming properties."

Incledon is a big fan of blood panels. In addition to screening patients for antibodies that might signal food intolerances, he looks at their vitamin and mineral levels and searches for evidence of fungal infections and environmental toxins. Because professional athletes are often pushing themselves to the edge of overtraining if not over it, he says, they are vulnerable to pathogens a less-stressed immune system would fight off without incident. In 2003 John Welbourn, a six-foot-five offensive lineman, had just been traded from the Philadelphia Eagles to the Kansas City Chiefs when he started having a range of problems, from extreme fatigue to thinning hair. A friend told him to go see Incledon. "Tom took like sixty vials of blood off me," Welbourn, who now runs a training company called Power Athlete, recalls. When the results came back, Incledon told Welbourn he had tested positive for several species of toxic mold. Welbourn believes he contracted it in the hotel where the Chiefs — "the cheapest fucking organization on the planet" — put him up while he looked for housing. Incledon put Welbourn on vitamins and supplements to build up his immune system and made some other esoteric recommendations, like telling him to avoid red and yellow foods. "It was literally like in *Lord of the Rings* when Gandalf casts out evil and the dude comes back to life within seconds," Welbourn recalls. (By "the dude" I know he means King Théoden in *The Two Towers*.) Ever since then, he has gone to see Incledon twice a year. "I realized conventional medicine, what doctors were offering me, was useless," he says.

Incledon first started to question the orthodoxies of sports nutrition when he was earning his undergraduate degree at Penn State in the early 1990s. An Olympic hopeful in weightlifting, he spent most of his nonclassroom time at the gym, where the serious weightlifters were guzzling protein shakes and scarfing cans of tuna. When Incledon asked his professors how much protein he should be consuming, they cited the federal recommended daily allowance guidelines — which, as noted above, are now recognized as inadequate for even moderately active exercisers, let alone competitive powerlifters. "I just looked at my professors and, I didn't want to be judgmental, but these guys are wimps, and they're telling me that I don't need more protein," he says. "And then I would go to the gym, and I see all these Neanderthals that are dumb as rocks, but they're fucking huge." Incledon chose to listen to the Neanderthals.

Speaking of getting huge, another treatment option Incledon offers clients is testosterone-replacement therapy. This raises some interesting questions, because "testosterone replacement therapy" is another way of saying steroids. In theory, it's "replacement therapy" if the person taking testosterone-boosting hormones has low T levels to begin with, but that nice distinction is mostly lost on sport-governing bodies like the World Anti-Doping Agency, the International Olympic Committee, and the NFL, all of which virtually never grant therapeutic-use exemptions for testosterone. To qualify for an exemption for any substance on WADA's prohibited list, an athlete must demonstrate that he/she would suffer "a significant impairment to health" in its absence, would experience "no additional enhancement of performance other than that which might be anticipated by a return to a state of normal health" while using it, and that there is "no reasonable therapeutic alternative."

It's impossible to talk about how elite older athletes are using sports science to prolong their careers without doping coming up — again, and again, and again. Any athlete in any sport who continues to perform at the highest levels after 30 automatically comes under suspicion of doing so only with the aid of performance-enhancing drugs, or PEDs, whether that means testosterone, human growth hormone, EPO, meldonium, or something more exotic.

There's a reason the shady doctor at the center of practically every doping scandal always seems to work at an "antiaging clinic." In a meaningful way, the point of doping is to make bodies function like they're young: fast-healing, fast-recovering, able to put on muscle easily and keep it on. It doesn't help that the most notorious dopers, like Lance Armstrong, Marion Jones, and Barry Bonds, were the ones who swore most indignantly that their late-career surges were 100 percent God-given.

After the truth comes to light, the excuse is usually something along the lines of: everyone else is doing it; I was doing what I needed to level the playing field. This is not true. Several of the athletes I interviewed for this book expressed anger, resentment, and moral outrage at the cheaters in their sports in a way that convinced me they meant it. I happened to be in Victoria for interviews with Hilary Stellingwerff the weekend news broke that Abeba Aregawi, the 2013 champion in the 1,500 meters, had been suspended after testing positive for meldonium, a drug used to treat heart conditions that also seems to improve the body's ability to use carbohydrates as fuel. Stellingwerff also runs the 1,500 meters. She had a baby in 2014 and was hoping to get pregnant again after the 2016 Olympics; taking any sort of drug with unknown health risks was out of the question to her. It absolutely infuriated her that she had to compete with people who barely tried to hide their PED use. I think there are more Hilary Stellingwerffs than Lance Armstrongs.

That said: yes, doping is widespread, and many of the athletes who want you to think they've turned back the hands of time through hard work, clean eating, or fringy New Age remedies are quietly getting a lot of help from pills, creams, gels, and syringes. But. No one *just* dopes. The Lance Armstrongs and Alex Rodriguezes may be lying about what they're not doing, but they're also doing all the things they want you to know they're doing, and they're doing them because they believe that those other things work, too. They may not work as effortlessly and dramatically as PEDs. Hell, they may not work at all; a lot of the interventions elite athletes swear by as longevity-bestowing are fancy placebos. The point is, the existence of doping doesn't negate all that.

Incledon, for his part, makes no attempt to hide the fact that he thinks steroids should be legal, if not in all sports then at least in football. A former member of the league's scientific advisory committee, he views the NFL's position that regulating PEDs is a matter of player safety as the height of hypocrisy. "You can't testify you're worried about safety when you're gonna let a man hit another human being as hard as they can, because that is not safe," he says. "And anyone that accepts that, they're a liar. You can bring in any five-year-old and ask him, show him a video, 'Does that look safe?' and they'll say no. So that's how you know a lot of the drug testing is a joke." If the NFL were serious about player safety, he says, it would have cracked down long ago on abuse of Toradol, a powerful anti-inflammatory painkiller team doctors inject into players to allow them to play through their injuries. "Toradol causes all kinds of ulcers, it causes all kinds of stuff, especially at the abusive levels NFL docs use it. Yet they'll tell a guy, 'Don't take growth hormone or testosterone because it's not safe.'"

In private, Incledon claims, NFL team owners have admitted to him the last thing they want is for a player they sign to go clean. "Whatever a guy was taking before they started paying him, they wanted them to take at least double now that the money's coming out of their pocket. No one wants to pay for weaker or lower goods. And then publicly, they would say, 'We're very confident no one in the team was taking any banned substances.' The team owners knew what's going on. They won't pay ten million, twenty million a year for a guy that doesn't know this stuff."

At the same time, he's skeptical of efforts to police steroid usage by monitoring for suspiciously high levels of hormones on the assumption they must be achieved artificially. The antidoping regulators, in his view, are making the same mistake his exercise physiology professors back at Penn State did when he asked them about protein needs. They're using reference ranges derived from studies of large populations rather than comparing athletes to other athletes. "You're basically saying, 'These guys are unique. They're so unique that only a few get to play at that level. But we're gonna hold them to the level of the average guy, and then say that's fair, when we don't really know if that's fair or not.'" This is an argument various players' unions have used in resisting drug-testing regimens. (It's also the argument you'd use to justify doing testosterone "replacement" on someone whose testosterone is within the normal range to begin with.)

In fact, he says, there's a strong case to be made that many football players may have legitimate medical reasons for some forms of doping. In 2002 Incledon sent a letter to leadership of the NFL Players Association. Incledon had noticed data suggesting that players who had suffered concussions had unusually low levels of human growth hormone. This was years before the long-term effects of concussions, including the devastating brain syndrome now known as chronic traumatic encephalopathy (CTE), became an issue of primary concern to either the NFL's management or its labor. Unlike testosterone, HGH's value as a performance enhancer isn't cut-and-dried: studies suggest it increases muscle mass but not force production. A deficiency of HGH, meanwhile, is associated with symptoms from decreased bone density to fatigue and depression. Incledon never heard back from the NFLPA. But he was on to something. Over the next few years, scientists published evidence of concussion-related endocrine dysfunction in kickboxers, soccer players, and soldiers with traumatic brain injuries from roadside bombs. In 2014 the NFL finally began testing players for HGH — not to see if they had enough, but to catch and punish those who were supplementing. "You can't criticize an athlete for trying to address a problem that the organization was ignoring, or saying it didn't exist," Incledon says.

As for testosterone, Incledon is a strong proponent but — he claims — a conservative clinician. He breaks with the medical establishment on two points: First, he disagrees with the view that testosterone levels on the low end of the normal distribution are not grounds for treatment. He likens it to IQ: a below-average number may not be indicative of a disorder, but higher is better. Second, he challenges the widely held view that testosterone production decreases with age in healthy men. "I'm not disputing that's what they showed in the data," he says. "What I'm disputing is their interpretation of what they're seeing." It could be that testosterone drops after 40 because after that age most men exercise less, or have less sex. In fact, a 2014 article in the journal *PLOS ONE* supported this claim, finding that, while variance between men in the amount of testosterone they produce goes up after 40, average levels remain the same.

But while Incledon isn't shy about saying more is better, he insists his preferred method of increasing testosterone is through nutrition, exercise, and lifestyle. It's more common, he says, for patients to show up at his clinic already taking supplemental testosterone and go off it with his help than it is for him to put them on it. I'm not sure how much I believe this. I'd certainly like to, because Incledon is, frankly, a really nice guy who, after six hours of nonstop interviews and tours, insists I join him, his delightful wife, and their daughter for dinner and wine. Then again, if there's one thing I've learned from my study of sports nutrition, it's that you can believe almost anything if you want to badly enough.

10

THE WORKOUT AFTER
THE WORKOUT

*Why Older Athletes Are Behind
the Recovery Revolution*

The coldest outdoor temperature ever measured on earth was −128.6°F, recorded at a Soviet weather station in Antarctica on July 21, 1983. For the past 90 seconds, I've been standing in a tiny room that's nearly 100 degrees colder. In my underwear. Listening to Beyoncé.

It's pretty cold in here. In fact, as a chill mist eddies around my bare legs and torso, it feels like it's getting colder. That's a trick of perception; a computer makes sure the flow of liquid nitrogen keeps the temperature in my shower-stall-size chamber steady at a bracing −220°F. Through a small window of triple-thick glass, I can just make out a seating area where a line of bathrobe-clad health enthusiasts wait their turns to take my place. I'm glad they're out here, because I'm starting to get a little worried. I was only supposed to be in here for two minutes, but it feels like maybe it's been longer than that already. Could the timer be broken? Is this version of "Crazy in Love" some kind of extended cut? Is someone going to let me out of here soon?

I'm at the offices of Cryohealthcare Inc., a clinic in Beverly Hills that offers a treatment called whole-body cryotherapy (WBCT) to customers seeking relief from a host of different ailments, from wrinkles and sore muscles to arthritis and postsurgical pain. On a busy day, which this weekend morning is already shaping up to be, up to 100 people troop through these offices for a $65 treatment, among them a number of well-known NBA stars, mixed martial artists, and Hollywood actors. Some of the bigger celebrities make after-hours appointments to avoid being seen. The boxer Floyd Mayweather came in here once with members of his entourage, The Money Team, and liked his two sessions in the cryochamber so much, he tried to go back in for a third; the staff had to delicately explain to the reigning middleweight champion that the protocol dictates only two exposures for first-timers. The clinic has become so popular, at the time of my visit the owners are scrambling to open a second branch in Woodland Hills and scouting real estate in Manhattan Beach for a third one.

Cryohealthcare is run by a German-born physician named Jonas Kuehne, his wife, Emilia, and his brother, Robin. Jonas first encountered WBCT in Europe during his medical residency; Emilia learned about it separately while working as a journalist in Germany. Together, they saw an opportunity to introduce it to the States, where it was virtually nonexistent, and purchased a unit for his family medical practice, which catered primarily to senior citizens on Medicare. "We were the first center offering whole-body cryotherapy in California," Emilia told me. "There are a lot of places claiming they were the first, but we were actually the first."

WBCT was developed in the 1970s as a remedy for rheumatoid arthritis. Athletes in Japan and Europe have been using it to treat postexercise muscle and joint soreness for decades, but it was relatively unknown in America before the early 2000s. Among the only athletes here hip to the (alleged) benefits of WBCT were professional basketball players, many of whom had tried it during stints in Europe. In 2009 a massage therapist who worked on several NBA players started referring them to Kuehne's clinic for cryotherapy. Word of mouth spread from there. Enough teams started inquiring about buying their own units that the Kuehnes, who had shuttered Jonas's medical practice in favor of offering cryotherapy full-time, established a second company to sell them. The LA Clippers, San Antonio Spurs, and Toronto Raptors have all bought units. A walk-in unit costs $98,000; a smaller one that chills only from the neck down runs $49,000. Kobe Bryant has a cryotank in his house, naturally.

For pro athletes, two or three minutes in the "cryosauna" is a more palatable alternative to the traditional postworkout ritual of soaking in a tub of ice water. In addition to being less time-intensive, it's easier to tolerate. "If you've done ice baths before, you're going to love whole-body cryotherapy," Emilia assures me as she hands me a robe, slippers, earmuffs, and a surgical mask and steers me into a changing room. Once inside the cryosauna, a black box about the size of a walk-in closet, I understand what she means. Plunging any part of your body into ice water, much less the whole thing, is sufficiently unpleasant that researchers studying things like pain tolerance and willpower use it as a form of safe torture. I've taken enough ice baths over the years that I've grown to enjoy them in the same perverse way I like spicy food or a nausea-inducing workout, but that moment when the first touch of ice water steals my breath is never fun. There's almost none of that in the cryosauna, just a gradually intensifying awareness that *gosh, this is actually pretty cold, isn't it?* By the time you're wondering if it might be too cold, it's time to get out.

But physiologically, ice baths and cryosaunas work in distinct ways, says Jonas Kuehne. You can even observe the difference with the naked eye. "In an ice bath, you see all the blood gets rushed to the extremity to warm it so you get some redness right away," he explains. "At the cryogenic temperature, the body is very smart. So you see the first behavior is to shunt blood away from the extremities to preserve core temperature," turning hands and feet pale, not red. At a cellular level, he continues, temperatures below -190°F stimulate the release of anti-inflammatory cytokines and suppress the production of pro-inflammatory tumor necrosis factor alpha (TNF) and C-reactive protein. To experience the full benefits, Kuehne says, takes 6 to 10 treatments.

In Europe, cryotherapy is used as an alternative to drugs for treating low-back pain, postsurgical pain, and other complaints. For cryotherapy providers in the U.S. to make any claims about its efficacy as a treatment for medical conditions, they would have to get approval from the FDA, which hasn't happened yet, but Kuehne does tell me, pointedly, that orthopedic surgeons in LA have started referring postoperative patients to his clinic to manage their pain and swelling without potentially addictive opioids like Percocet or Vicodin. "We can't make official statements saying we're using it to treat X and Y," Kuehne says. "The U.S. is very particular in only allowing U.S. studies — but there are so many studies already out." This I'm-not-saying-I'm-just-saying act wasn't quite subtle enough for the FDA, which in August 2016 sent Cryohealthcare a stern warning letter telling Kuehne to stop making quasi-medical-efficacy claims or face legal consequences.

This being Beverly Hills, of course, a lot of the clients who come through are looking at it less as a trip to the doctor than as a spa day. Cryotherapy supposedly tightens up the collagen in skin, and the whole-body variant is said to burn more than 500 calories at a pop. "A lot of clients need it for pain management, but they're more fascinated with the beauty aspect," says Emilia. The Kuehnes themselves make quite a good advertisement for that aspect: none takes fewer than three treatments a week, and all three look like models from a high-end ski-resort brochure.

Jonas Kuehne, who used to do a fair amount of kickboxing, says he enjoys working with his athlete population most. "I like them because they know very well how their bodies respond to anything. These guys don't waste their time. They wouldn't do a treatment that, maybe it helps, maybe it doesn't."

In fact, while there are plenty of scattered data points in the published literature to suggest whole-body cryotherapy does something to accelerate recovery from exercise, the evidence falls somewhat short of the considerable hype. A 2010 literature review in the journal *Sports Medicine* mostly supported Kuehne's claims about the effects of cryogenic temperatures on molecular inflammation pathways and said it "should be considered a procedure that facilitates athletes' recovery." But the authors hedged that conclusion, noting that few high-quality studies on WBCT exist. And they specifically challenged Kuehne's contention that the effects build up over the course of repeated treatments. In fact, they said, the body seems to mount an adaptive response, suggesting frequent treatments would have diminishing returns. In individual studies with runners and other athletes, WBCT treatments have been associated with a modest acceleration of "functional recovery" as measured by tests of strength and flexibility and a decrease in self-reported soreness. But again, there haven't been many such studies, and the ones that exist tend to incorporate smallish sample sizes and less-than-ideal experimental design.

To the extent there's a scientific consensus on WBCT, it's that doing it after exercise will help you recover faster than if you do nothing, but so will a lot of things that don't cost $65 a pop. And money isn't the only potential downside. In 2015 an employee at a Las Vegas spa was found dead in a cryosauna by her coworkers when they showed up to work. Her body was frozen solid. Investigators speculated that she had dropped her phone on the floor of the unit while enjoying a free session. When she bent down to retrieve it, the oxygen-poor atmosphere at ground level caused her to pass out. That was, of course, a freak accident. The safety precautions at a place like Cryohealthcare preclude worry about even mild frostbite, much less death. But if you're looking for another reason to save yourself $65, there's that.

I didn't work out before my cryosauna session, so I can't say whether it had any effect on my recovery or soreness. What I did notice was something a lot of WBCT devotees report: a pleasant buzz of mild euphoria and energy that lasted several hours afterward. That could be because I had Beyoncé in my head. Individual results may vary.

The modern athlete spends a lot of time in boxes. And not just boxes but tubes, tubs, tanks, tents — containers of all shapes and sizes. When he's not in some sort of container, or sometimes when he is, he's also likely to be strapped down, plugged in, or hooked up to some sort of smaller box that emits laser beams or ultrasonic waves or electrical current. Most of these containers and devices are aimed in one way or another — sometimes in ways that seem downright contradictory — at promoting recovery.

There are hot tubs and cold tubs. There are cryotherapy chambers, which claim to supercool the skin without chilling the core, and infrared saunas, which use ultra-low-frequency light waves to warm the core without toasting the skin. There's contrast hydrotherapy, which uses alternating blasts of cold and hot water to suppress inflammation and increase circulation — in theory, the best of both worlds. There are hyperbaric chambers, booths or tents in which athletes breathe oxygen at several times the pressure of atmosphere at sea level. There's something called the NormaTec, which is sort of like an inflatable pair of pants that uses rolling waves of pressure to flush lactic acid and other metabolic by-products out of muscles. There are so-called float tanks, filled with saltwater with a specific gravity lower than the human body; lie in one with the lights off and you'll feel like you're drifting disembodied through an infinite void. Some people, like Golden State Warriors guard Stephen Curry, say there's no faster way to recharge your brain. Others find it brings on panic attacks. (Pro tip: don't get in one right after a close shave. Water that salty *burns.*)

There's even something called vinotherapy, which is nothing more or less than taking a bath in a tub full of hot red wine. As far as I know, there is only one vinotherapy facility in the U.S., at a luxury spa called Aire Ancient Baths in lower Manhattan. When Amar'e Stoudemire was playing for the New York Knicks, he got a vinotherapy session as a Father's Day gift. "I had never bathed in red wine before and I thought it was a pretty good thing to do," he told me. Stoudemire became a regular at Aire, soaking his 6-foot-11 frame in a copper tub full of Spanish tempranillo after home games. "It helps reduce pain and it also gives your red blood cells a boost as far as your healing process is concerned," he says, somewhat uncertainly. Pressed further on its benefits, Stoudemire shrugs. "I don't feel like any magic happens as soon as I step out of the tub. I feel like after a regular hot bath." Despite that lukewarm endorsement, I was determined to try it for myself. Alas, when I stopped by Aire on a visit to New York, the vinotherapy suite was being renovated. I spent an hour soaking away my disappointment in splendid marble baths of various temperatures and salinities, with impressive Latin names: the steamy caldarium, the icy frigidarium, the salty flotarium. Afterward, the manager, a Spaniard named Fernando, told me about some of his other famous clients, including Rafael Nadal, who, he said, comes in every other day during the U.S. Open.

If there's been one unignorable trend in sport science in the last few years, it's the newfound appreciation of recovery as a pillar of athletic performance, on par with conditioning and nutrition. Ralph Reiff, executive director of St. Vincent Sports Performance in Indianapolis and the former head of athlete care for the 1996 Olympic Games in Atlanta, calls it "the next frontier in human performance." He estimates that 95 percent of the research informing the new understanding of recovery has been done in the last 10 years. "If you were to ask me the question, 'Hey Ralph, how much more are we paying attention to that over the last decade than we did previously?' I would say we never paid attention to it previously," Reiff says.

That's certainly not the case anymore. These days, "recovery" is probably the hottest buzzword in sports marketing. There are multiple fitness club chains that offer nothing but stretching classes and a popular fitness program called MELT Method built around self-massage with a foam roller. The San Francisco float spa Steph Curry frequents just opened a sister location, with a cryosauna, a mile from my house in Berkeley. Gatorade now sells postworkout shakes with a blend of carbohydrates and protein — "exactly what your body needs to recover and rebuild." (Never mind that what Gatorade's selling is basically just expensive chocolate milk, the beverage that, according to nutritionists, features those nutrients in the optimal blend for muscle repair.) UnderArmour makes a skintight compression garment meant to be donned after competition; wearing it for 24 hours (!), the company claims, yields benefits including "significant reduction in muscle damage quantified by levels of creatine kinase in the blood; a reduction in subjective fatigue level by 50 percent; a reduction in subjective muscle soreness by 50 percent." (Studies have suggested that compression garments can indeed reduce soreness and speed functional recovery, but they have to be uncomfortably tight to work.)

UnderArmour also partnered with Tom Brady on a line of "recovery sleepwear." The garments are lined with a "soft bioceramic print" that absorbs body heat and reemits it as "far infrared" radiation. Basically, it's an infrared sauna in pajama form. That's the idea, anyway. As evidence that its product is informed by real science, UnderArmour points to a 2012 study in the journal *Photonics & Lasers in Medicine*. But the authors of that study, while noting that far-infrared radiation has demonstrated effects in animal and in vitro cell studies, simply note it's theoretically possible to create a fabric that delivers similar benefits for humans. Basically, they said, wouldn't it be cool if someone could prove this works? and UnderArmour cited that as evidence it works.

Science or pseudoscience, Brady is the perfect face for the recovery trend, which has particular appeal for older athletes. As we've seen elsewhere in this book, in most sports, it's not the decrease of maximum muscle contractile force or blood-oxygen capacity that limits athletes' ability to compete with age. More often, it's the need for ever-longer breaks between performances to regain strength and range of motion and the increasing propensity to injury as fatigue piles up. Brett Favre, the Green Bay Packers quarterback who holds the NFL record for consecutive games started, told me it wasn't a loss of ability but the difficulty of bouncing back in between games that pushed him into retirement after 20 seasons. "It wasn't an overnight deal but I woke up one day, at least it seems like, and I was like, 'You know, I just don't recover like I used to,'" he says. In his 20s, Favre needed only a day or two to recuperate from the punishment he took on Sundays. In the last five years, he says, he would make it to the following weekend still not sure if he'd be able to play. "Then, about game time, you start feeling better. And then you start back over again." It's a familiar story.

Cryosaunas, hypberbaric pods, and the like are a direct answer to this problem and they're an answer that doesn't make the problem worse. That's crucial because athletes are control freaks. As they feel themselves losing explosiveness, the temptation is to respond by ramping up training, which can easily result in overtraining. "One of the hardest things to do for an elite athlete is take a day off," Meb Keflezighi says. "We're so motivated, so driven by routine. We like to go, go, go because we want the rhythm."

As Keflezighi neared 40, he channeled that nagging drive into his recovery regimen, which included stretching, massage therapy, self-massage, and the NormaTec pressure sleeves. Keflezighi went from taking ice baths, which he hated, to using something called the Hyperice, a sort of compression bandage with a built-in ice pack. "Ten percent of my job is running. Ninety percent is not running," he said as a competitor. "The running is the easy part." Even on one of his long-run days, he'd only be on the road for three hours, he pointed out. "The next twenty-one hours, how you use it is crucial."

And this is what explains the recovery boom as much as anything. If you keep adding more and more exercises onto your workout, very quickly you will pay a price for overdoing it. With recovery, there's no such thing as too much. You could spend every waking hour that you're not training shuttling between the hot tub and the cold tub and the worst that'll happen is you get pruney fingers.

Actually, you can spend more than every waking hour. Sleep is the most important type of recovery, and plenty of athletes try to make the most of it, whether with compression suits and infrared pajamas or more high-tech tools. T.J. Oshie, a winger for the Washington Capitals, goes to bed every night wearing adhesive gel pads connected by wires to a device called ARP, for Accelerated Recovery Performance. The electrical impulses it sends through his muscles cause them to contract and relax, stimulating blood flow that's supposed to promote regeneration and healing. ARP isn't much different from the TENS (transcutaneous electrical nerve stimulator) Shaquille O'Neal markets on late-night TV or the Japanese-made Sixpad Body Revolution, an electrical muscle stimulator (EMS) advertised by Real Madrid star Cristiano Ronaldo. EMS exercisers may sound like a joke but studies have shown they do produce modestly stronger muscles. They're not going to make you look like Ronaldo, but your physiotherapist might have you use one while you're rehabbing an injury and unable to do other exercise. Likewise, there's some research showing ARP does what it's supposed to, but probably not enough to justify wearing sticky pads on your body 12 hours a day, like Oshie does.

When you get down to it, that's the story with most popular recovery modalities. Some might work a little better for soreness while others are better for restoring strength or range of motion; most of them are more effective than doing nothing ("passive recovery") but not by as much as you might expect. It's hard to make any categorical statements, as there just hasn't been that much high-quality research and the effect sizes observed are usually small, especially when it comes to alternative medicine practices like cupping, in which heated cups placed on the skin create a vacuum effect that supposedly stimulates blood flow. (Michael Phelps received a number of cupping treatments in Rio, creating a sensation when TV cameras noticed the large, round bruises on his back.) Where there is more data to rely on, it tends to weaken rather than strengthen the argument for interventions widely regarded as useful. Massage therapists are ubiquitous in professional and Olympic training rooms, but a meta-analysis of 22 studies looking at the effects of massage on athletic recovery found surprisingly little support for postworkout rubdowns. The authors noted some evidence that it helps with short-term recovery from high-intensity mixed exercise but concluded, "It remains questionable if the limited effects justify the widespread use of massage as a recovery intervention in competitive athletes."

Balanced against the modest benefits of recovery interventions are the real costs. Even if your budget allows for daily cryotherapy sessions, they might not be a great idea. No, there's no such thing as "overrecovery syndrome." You're not going to throw your immune and endocrine systems out of whack the way you can from overtraining. But just as taking superdoses of antioxidants or anti-inflammatory medications can blunt the body's adaptation to training stimulus, recovery treatments aimed at suppressing the inflammation and soreness that are a natural response to exercise can do much the same thing. If feeling good is your primary goal, then by all means spend as much time hopping from ice tub to hot tub and back again as you can. But if you're looking to improve performance, the consensus is reserve intensive recovery treatments for periods of competition, not training.

And there may be some treatments better off avoided altogether. Hyperbaric oxygen chambers are increasingly popular with athletes like Novak Djokovic, who sits in a pressurized chamber and watches movies after his matches, and NFL running back Rashad Jennings, who paid $18,000 for a home unit so he can sleep in it. Joe Namath, the retired New York Jets quarterback, goes even further; he believes subjecting himself to more than 120 hyperbaric sessions reversed the cognitive decline he was suffering as a result of repeated concussions, and pledged to raise $10 million to fund the studies to prove it. Unfortunately, the evidence that hyperbaric therapy does much of anything, let alone cures CTE, is largely lacking. One study of 24 subjects found it didn't accelerate return of strength or dissipation of soreness after exercise. Another noted greater "perceived recovery" but attributed it to a placebo effect. Meanwhile, at least one researcher has speculated that breathing pure oxygen at high concentrations could promote cancer growth.

There is one recovery intervention whose benefits are undisputed: sleep. It's during deep sleep that the body's production of growth hormone peaks, and accumulated sleep debt is associated with depressed protein synthesis and sarcopenia, or muscle wasting. Unfortunately, sleep quantity and quality both have a nasty way of declining steadily with age: the average 75-year-old gets 90 minutes less per night than the average 25-year-old. The decrease is thought to be caused by a gradual dying-off of neurons in a region of the brain called the ventrolateral preoptic nucleus. Scientists believe that cluster of cells regulates and organizes sleep, suppressing impulses that might interrupt it; in animal studies, rats with smaller VPNs were observed to suffer from "sleep fragmentation." To make up for the lack of quality sleep, many (human) athletes who continue to excel in their 30s and 40s commit to spending extra hours in bed. Roger Federer, LeBron James, and the marathoner Deena Kastor are among those who say they aim for 12 hours a day.

For most of us, spending half of every day lying down is impossible, and those of us with demanding jobs that don't involve playing games and exercising are lucky to grab even eight. Like T. J. Oshie with the ARP machine and Rashad Jennings with his hyperbaric pod, many athletes are willing to try almost anything to make those hours count even more. Aaron Rodgers, the Green Bay Packers quarterback and two-time NFL MVP, struggled to get even seven hours of sleep until he cut out afternoon coffee and began taking a bedtime supplement containing zinc, magnesium, and aspartate, said to be a natural muscle relaxer. Rodgers is also a sometime practitioner of Earthing, an alternative practice based on the belief that "connecting to the Earth's natural, negative surface charge" by sleeping directly on the ground or on a special conductive mat stimulates blood flow and healing, according to something called the Earthing Institute. Unlike most alternative practices, Earthing is patented and trademarked. Of the handful of published studies examining the effects of Earthing, or grounding, as it's also called, almost all were sponsored by a company called EarthFX, which sells products like the $260 Earthing Recovery sleeping bag, and which appears to operate the Earthing Institute website as well.

Perhaps the most extreme recovery regimen I've come across belongs to Phil Wagner, the founder of Sparta Science, a Silicon Valley performance center that uses force-plate jump testing to shape training programs for athletes. Between running his company and raising his three small kids, Wagner allots only four hours per night for sleeping. To wring the most out of it, he uses a biofeedback device called the EmWave at bedtime to regulate his heart-rate variability. HRV is basically a measure of how well the heart responds to changing demands on the body, speeding up and slowing down as needed. A low HRV score indicates a fatigued, unresponsive autonomic nervous system; a high score, one that's recovering well. An increasing number of coaches use HRV monitoring systems as a guide to when to push their athletes and when to give them a rest day. The technique Wagner performs using the EmWave, called coherence training, is said to produce a state in which the sympathetic and parasympathetic nervous systems work together to produce smooth waves of rising and falling heart-rate variability. Wagner, a former Division I safety and semipro rugby player, claims his coherence training allows him to spend more time in restorative deep sleep for every hour he's in bed. When he rises, he spends the first few minutes of the day sitting under red lights to stimulate the mitochondria in his retinas to ramp up their energy production. Wagner also takes cold showers every night and fasts 18 hours out of every 24, citing studies showing that "intermittent fasting" promotes cognitive clarity. As a recovery program, that sounds pretty exhausting.

11

REPAIR, REPLACE, REJUVENATE

Sports Surgery and the Outer
Limits of Human Longevity

At the far eastern end of East 70th Street on Manhattan's Upper East Side, on a skyway connecting two office buildings, is a large sign: HOSPITAL FOR SPECIAL SURGERY: WHERE THE WORLD COMES TO GET BACK IN THE GAME. Inside the building on the uptown side of the skyway, Dr. Bryan Kelly is explaining to me what went wrong with Alex Rodriguez's hips.

On his computer monitor, Kelly, a beefy man in his 40s, shows me a 3-D rendering of the hip joint — the same one, more or less, he showed to the New York Yankees third baseman when he sat where I'm sitting three years earlier. With the numerous strappings of muscle, tendon, and ligament digitally stripped away, the hip is a deceptively simple mechanism, just a ball swiveling in a socket. The ball is the head of the femur, which resembles the knob on the end of a walking stick, if the top few inches of the stick were cocked at a 45-degree angle. The socket is a bowl-shaped hollow in the underside of the pelvis, lined with a bed of spongy articular cartilage and rimmed by a band of tougher fibrous cartilage called the labrum, which rims the bowl like a rubber gasket on the lid of a mason jar. Quite often, the ball isn't a nice, smooth, geometric sphere like the leather-cork-and-rubber baseballs Rodriguez hit over outfield fences 696 times during his 22-year career in Major League Baseball, good for fourth on the all-time list (with an asterisk). It can be more irregularly shaped, like a snowball or a meatball. It's not totally clear why this happens, but it seems to involve both genetics and developmental factors. A hip joint subjected to repetitive stresses before the age of 15 — like, say, swinging a baseball bat hundreds of thousands of times — is more likely to become deformed in response to those forces.

"A femoral head that's not spherical has a prominent area of bone in the front, and that area of bone is what impinges when you flex and you rotate," Kelly explains, "so as you flex beyond ninety degrees and rotate, the labrum" — the rubber-gasket part — "gets compressed against the area of impingement on the femur. Eventually, with the repetitive impacts from this bump of bone, the labrum starts to detach from the socket and it loses its connection to the bone and also to the articular cartilage, and then it becomes unstable and it gets stuck in the joint like a hangnail, and then eventually this cartilage inside the joint can wear down." Then he puts it in less technical terms: "It's like a cheese grater that's grinding away the cartilage inside your joint." I wince as he says this, thinking of the impingement in my own right hip, diagnosed by MRI exam five years earlier. It hasn't caused me too much trouble since I made a few modifications to my workout to avoid aggravating it, but the thought of a cheese grater gnawing at my cartilage is still plenty disturbing.

It gets worse because in this case, the cheese grater itself grows bigger the more it grinds away, as the chronic friction causes calcium to build up on the protrusion. For Rodriguez, the buildup eventually became severe enough to require surgery in both hips. By the time he had his left hip done, in 2013, the bony outgrowth was limiting his range of motion so much he couldn't lift his knee an inch off the examination table. The surgery to restore normal function began with Kelly making an incision about the diameter of a dime in Rodriguez's groin and inserting an arthroscope, a tiny camera on the end of a tube. Using the image from the arthroscope, he went in through other small incisions, trimmed away the "hangnail" portions of Rodriguez's labrum, sutured the remaining portion back to the acetabular cartilage, and shaved away the bony protrusion on the femoral head, restoring its spherical shape.

Four months later, Rodriguez was back on a baseball field, running and hitting off a tee. He only managed to play in 44 games that season, and, at 37, was clearly no longer the player he'd been in 2009 when, coming off his first labral repair, he'd batted .286 and finished tied for 10th in MVP voting. A cynic might say his decline had something to do with the discontinuation of his steroid use, the revelation of which earned him a suspension for the entire 2014 season and that asterisk on all his records. But that he was able to play at all after 2009 is something of a medical triumph. Had Rodriguez been born a decade earlier, his career most likely would have ended then, seven seasons and 143 home runs sooner.

If there's a living symbol of how far orthopedic sports surgery has advanced over the last generation or so, it has to be Adrian Peterson, the Minnesota Vikings running back who led the NFL in rushing and came within nine yards of breaking the single-season all-time rushing record in 2012 after undergoing reconstruction of his ACL and LCL the previous December. In terms of player-games lost, torn ACLs are the most devastating injury in football. The anterior cruciate ligament, so named because it crosses diagonally over the front of the knee behind the kneecap, provides lateral stability to the leg during changes of direction. Straight-line athletes like sprinters or cyclists seldom rupture their ACLs; in football, they're an epidemic, ending 48 players' seasons in 2015 alone. Although most players are medically cleared to play again six to nine months after ACL reconstruction, it's often said to be a two-year injury, since that's how long it typically takes to return to a pre-injury performance level. Even more than his size or speed, what distinguished Peterson's game was his unmatched ability to make hard lateral cuts, bouncing runs out to the side when his running lanes got clogged up. Yet there he was in 2012, barely 10 months after surgery, looking, to the naked eye and to the statisticians, better than ever. James Andrews, the prominent sports surgeon who reconstructed Peterson's knee, also happened to perform one of the last procedures on Eric Dickerson, the running back whose 1984 record of 2,105 yards in a season Peterson just failed to break. Had modern arthroscopic technique existed when Dickerson played in the late 1970s and early '80s, he has said, Dickerson would be remembered as the undisputed greatest ball-carrier in NFL history.

Peterson's recovery was an outlier but it wasn't an anomaly. The days when a torn ACL was more likely than not a career ender — involving grisly open surgery and, more often than not, removal of the meniscus, which usually tears when the ACL does — are well in the past. Arthroscopy has been the preferred method for knee surgery since the 1980s. By the time Bryan Kelly was beginning his fellowship training in sports medicine in 2001, the ACL was no longer a puzzle. The technique had been pioneered; all that remained were incremental refinements. "Most of what we're doing for modern knee surgery was pretty much done by the nineties," he says.

For a would-be pioneer like Kelly, there was only one real frontier left in orthopedics: the hip. The biggest joint in the body was also the least understood. Buried deep within the core, it was like some ocean ecosystem around a volcanic vent five miles below the surface: scientists knew it was there, but its inaccessibility made it hard to study in situ. Compared with knee or shoulder injuries, the clinical presentation of hip problems is "a little more subtle," Kelly says, often masquerading as other injuries or manifesting in the form of compensatory pains elsewhere in the body. "A lot of people with what ended up being intra-articular hip injuries were being diagnosed with soft-tissue strains and pulls, like hip flexors and groin pulls," he says. "I think probably it was one of the reasons why you would probably see people's careers decline. They ended up not being able to perform at the level they needed to or they missed a lot of time." Not only had the technique he used on Rodriguez not been developed yet, it wasn't even generally accepted as a theoretical possibility. Just as complete removal of the meniscus had once been standard in ACL repairs, many orthopedists believed the labrum, with its minimal blood supply, was incapable of healing and the only justification for operating on it was to remove those stray "hangnail" bits so they wouldn't cause the joint to lock up. Meanwhile, the only way to get at the labrum was through on open procedure called a hip dislocation, which required a great deal of recovery time.

But beginning in the 1990s a Swiss research group led by a surgeon named Reinhold Ganz had been publishing papers advancing a new understanding of hip pathology and demonstrating the efficacy of labral repair combined with femoral-head resurfacing, all performed via arthroscopy to minimize collateral damage. The idea of specializing in an emerging area where he would have the opportunity to advance not just the application of a technique but the theory of it appealed to Kelly, so he moved to Pittsburgh to train under Marc Philippon, the surgeon who was then spearheading the adoption of Ganz's research in the U.S. It was Philippon who would perform Rodriguez's first hip surgery in 2009; he has also operated on the likes of pitcher Tim Lincecum and quarterback Kurt Warner. By the time Rodriguez went under the knife for the second time, labral repair with femoral-head resurfacing had gone from a semi-experimental procedure for athletes desperate to salvage their careers to an utterly routine one for any active person hoping to avoid a hip replacement. Kelly now does a couple of hundred a year. Most of those surgeries are performed on athletes under the age of 25, but there's no age limit per se. "You can operate on somebody in their sixties if their cartilage is healthy. The diminishing returns have got to do with the severity of the cartilage wear inside the joint. It just happens that cartilage tends to wear out over time.

"I'd say for the last seven years much of what I've done is pretty much the same," he adds.

And with that, there were no more worlds to conquer. The last joint stymying orthopedic surgeons had been solved. So what now?

An old medical joke: Man goes to the doctor. Doctor, he says, it hurts when I go like this.

So don't go like that, the doctor says. *Ba-dum-bump.*

One of the sharpest differences between elite athletes and the rest of us is how they approach surgery. A term you hear a lot from athletes and their doctors is "clean-out." At the end of a season of football or hockey or basketball, more players than not, it seems like, stop by the hospital for a quick one en route to the beach or golf course. It's a purposely vague word as professional teams are reluctant to share injury information that a competitor might be able to use and athletes don't want to announce anything that might hurt their market value. Usually "clean-out" denotes some minor arthroscopic procedure like removal of bone chips, spurs, or scar tissue causing pain or limiting range of motion, or debridement of the ragged edges of torn ligament or cartilage. Elite athletes get their joints serviced like the rest of us get our teeth cleaned.

That's not how it generally works for nonathletes. When you or I see the doctor about a new knee or shoulder pain, the response is almost always the same conservative protocol: First, stop doing the activity causing pain for a few weeks and see if it improves. If it doesn't, try a course of rehabilitative exercises for a few months, perhaps paired with a steroid injection to calm inflammation. A conservative orthopedist won't even order an MRI before all this has been tried, knowing that the correlation between abnormal MRI findings and pain or dysfunction is far from precise. Study after study has shown it's common for "healthy" people to have herniated spinal disks, meniscus tears, and other defects without any symptoms, and to have symptoms without observable defects. Promiscuous imaging, the thinking goes, ends up leading to surgeries that don't remedy the pain because they weren't the cause of it to begin with.

All that makes sense. But I've always wondered if it's the whole story. If conservative management is usually the medically appropriate option, then why isn't that how doctors treat professional athletes, who receive the best health care in the world because their health is worth millions of dollars to the people paying for it? In a world where the cost of health care was no issue, where our only job was to take care of our bodies as well as possible, would we all be stopping by the OR once a year for a quick intra-articular nip and tuck?

Yes and no, is the answer. One reason elite athletes undergo more surgeries and nonsurgical interventions than even highly active amateurs is because they are, in effect, living in the future. Lawrence "Rusty" Hofmann is an interventional radiologist at Stanford University and cofounder of a technology company called Grand Rounds that connects difficult medical cases with the most qualified 1 percent of expert doctors, at the cost of $7,500 per case. Hofmann came up with the idea for his company after his eight-year-old son developed aplastic anemia, a life-threatening bone marrow disease, and Hofmann was forced to scour his professional network to get the boy the cutting-edge bone marrow transplant needed to save his life. Making the case for Grand Rounds' services, he likes to cite research showing there's a 17-year gap between the scientific validation of new medical knowledge and the point when that knowledge becomes the basis of standard treatment. (In his PowerPoint pitch deck, Hofmann uses a slide of a 17-year cicada to drive this idea home.) Elite athletes, and the wealthy who can afford concierge services like Grand Rounds, cluster at one end of this 17-year timeline. The rest of us fall elsewhere on the spectrum depending on factors like how close you live to a major hospital, what kind of insurance you have, and who your doctors are.

Sports medicine practitioners with access to state-of-the-art tools and techniques are more aggressive than they would be if they didn't have those things. "When we look at surgeries for any athlete, and especially our older athletes, it's all about risk and reward," says Michael Terry, team doctor for the NHL's Chicago Blackhawks. "We're expanding our indications because risk and reward is tilting in that direction. The procedures are much less morbid than they used to be with a smaller downside and less risk to do some of these things."

Just as crucial as improvements in surgical technique to better outcomes have been advances in postsurgical rehabilitative protocols. Thirty years ago it was still common to cast joints after many surgeries in the belief that immobilizing them promoted faster healing. Not only did that turn out to be wrong, but the data has shown the opposite is true. Casting promotes stiffness of the joint and atrophy of the surrounding muscles. Newer rehab protocols have patients flexing the repaired joints starting on the first day after surgery, often with the aid of a device called a continuous-passive-motion machine.

"I think the surgery part of things is becoming less and less important," says Nirav Pandya, an orthopedic surgeon at the University of California at San Francisco who treats both youth and mature athletes. "The surgical techniques themselves have pretty much gotten to the point where you're not going to see much change in terms of the way we do things. I always tell my patients, 'I put this ACL in, my work's been an hour and a half, but what's going to determine the level of sport you get back to is how quickly you get that muscle strength back.' Those eight months after surgery — I think that's the next level of sports medicine."

Rehab is another reason surgeons are more inclined to do all those borderline procedures on elite athletes: they make great patients. Going into a surgery with a high fitness level and coming out of it with no requirements other than to focus on their rehab ensures much better outcomes than nonathletes can expect. When I asked Kerri Walsh Jennings, the Olympic beach-volleyball player, whether she had finished the rehab for her fourth and most recent shoulder surgery, she told me she intended to keep doing the program indefinitely, even though she was back to competing with no limitations. "I want to be playing volleyball when I'm ninety so I have to keep my shoulder strong," she said.

But the indications for nonathletes are also tilting toward surgery as advances in medicine — such as the elbow-ligament surgery named after baseball pitcher Tommy John — trickle down from elites to the general population. "Ten years ago, you would say, 'You're fifty, you're not getting your ACL done,'" Pandya says. "The outcomes are getting better because we're realizing that there are things that you can do that they actually have the potential to heal." But more effective treatments are only part of it. There's also vastly more demand from older patients with sports injuries or athletic limitations. "The age group that you see who considers themselves a sports medicine patient is a lot older now than that would even walk into a clinic fifteen years ago. The average age has probably gone from twenty to thirty. Suddenly you're seeing forty- and fifty-year-olds who want to get back to activities that, ten to fifteen years ago, you would have said, 'Wow. That's intense.' Downhill skiing, intense basketball, really competitive tennis, marathon running. The older patients have the expectations of younger people now."

But there's a difference between wanting to perform like a professional athlete in one's fifties and being treated like one. Those in the former camp may in fact wish to avoid the latter. All those clean-outs come with a price. There's a saying among surgeons: once you go into a joint, it's never the same. Recall that the body seems to have only a limited supply of the specialized stem cells that repair damaged tissues. Every surgery cuts into that supply a little more while leaving behind new scar tissue — tissue that itself might necessitate a clean-out down the road. "When you get over two, three surgeries, the chance of you getting better I think incrementally goes down, because you get stiffness, you get all these other issues," Pandya says. "Very rarely do you see an athlete who gets four knee surgeries who ends up being at the same level. If you get a clean-up, we know for sure you're going to be back. Over time, I think it probably does decrease the life span of these athletes. They're basically just wearing down their joints. Eventually, if you tinker so much on the Ferrari, it just stops working." (Did you really think we were going to get through this without a car metaphor?)

From the athletes' perspective, though, the conservative option has its own price. You can't tell a professional sports player "Don't go like this" when going like this is his livelihood. If you're earning $10 million a year, a surgery that adds another year or two to your career but raises your risk of debilitating arthritis in 10 years' time isn't necessarily a bad trade-off, just a depressing one.

The good news is that it's a trade-off athletes are less and less likely to have to make thanks to advances in biologics, a catch-all term for therapies that use living cells, often taken from the patient's own body. At the very moment sports medicine has reached the limit of what can be achieved with a scalpel and forceps, a new frontier has opened up offering the promise not only of repairing injured tissue but of restoring it to a true pre-injury state — rejuvenating it, in fact. "We've shifted a lot of our focus toward biological treatments for orthopedic problems," says Terry, the Blackhawks' team surgeon. "While the vast majority of things still have mechanical solutions as the tried-and-true, biologic solutions are going to provide the same outcomes and not alter anatomy. Things like platelet-rich plasma, stem cell treatments, gene therapy — those things are definitely on the horizon and will be important in the treatment of athletes and, really, everybody."

PRP therapy — the technique that uses a centrifuge to separate out platelets from blood plasma so they can be injected into injury sites to accelerate healing — has been viewed as a panacea by American pro athletes since Kobe Bryant was reported to have flown to Germany in 2012 for treatments to alleviate osteoarthritis in his right knee. (What Bryant got was actually Regenokine, a patented cousin of PRP in which the isolated plasma is heated in a way that supposedly primes the anti-inflammatory proteins in it.) PRP has shown enough promise in the treatment of muscle and tendon injuries that it's widely used as an adjunct to surgery, even though research to determine the most effective protocols is still under way. There's also somewhat weaker data showing it helps with arthritis symptoms. Critically, unlike surgery, "it has very low risk associated with it," says Terry. "It's essentially very noninvasive, as a treatment goes."

The other thing that makes biologics so exciting is they're the rare intervention that might actually be more effective for older patients than younger ones. That's because the way many of these treatments work is essentially by reproducing or mimicking the marvelous self-healing powers of a younger body, with its abundant stem cells and growth factors. "In the young kid you have such good healing potential" that something like PRP is probably redundant, says Pandya. "But take that person who may have had a couple of injuries in their knee when they played college sports, and now they're thirty-five or forty and it's just bothering them. The answer before was, 'Just stop.' Now, it may be, 'Let's grow some cartilage in this area. Let's see if we can get your body back to when you were twenty through some of the cell and molecular stuff that we're doing.' We're looking probably over the next five to ten years where a lot of that stuff will happen."

Cartilage, in particular, is a tissue reliably affected by age. About a quarter of all adults over 55 show signs of knee osteoarthritis, the joint inflammation that occurs when cartilage breaks down, and around 10 percent have symptoms. In competitive athletes, the numbers are higher. Attempts to regrow cartilage have a fraught history. For a period of time between 2005 and 2010, there was a great deal of excitement in the sports medicine world about a procedure called microfracture, in which tiny holes are drilled into the bones around the joint (generally the knee). The seepage of blood into those holes stimulates production of new cartilage, which seldom happens on its own in adults. Microfracture was especially popular with NBA players, with stars like Amar'e Stoudemire, Penny Hardaway, and Greg Oden all going in for it. But the long-term success rate of microfracture turned out to be poor. The type of cartilage produced isn't spongy articular cartilage but stiffer fibrous cartilage, and the resulting composite of cartilage types lacked structural integrity.

A higher quality of cartilage can be attained with a procedure called autologous chondrocyte implantation. It's actually two procedures. In the first, chondrocytes, or cartilage-producing adult stem cells, are harvested from the knee and isolated from the cartilage matrix so they can be cultured in the lab to increase their number. In the second, they're reimplanted in the knee underneath a flap put in to hold them in place. Compared with microfracture, ACI produces more durable tissue. Compared with another technique, osteochondral implantation, which takes cylinders of excess healthy cartilage from areas of the joint it's not needed and uses them to plug holes, it's able to fix larger defects. In three clinical trials of ACI involving a total of 183 athletes, 78 percent were able to return to their pre-injury level of play within 25 months. But that "within 25 months" part is the rub. To allow the new chondrocytes to take root, the patient must abstain from impact sports for up to 18 months. As we've seen, to keep their jobs, plenty of professional athletes will opt for a short-term fix that gets them back in action sooner.

Before long, they may have a better option. At the Scripps Clinic in La Jolla, California, a doctor of biophysics named Darryl D'Lima is perfecting a treatment that involves using a 3-D printer to apply a new matrix of living cartilage stem cells exactly where they're needed. While it will still require months for the progenitor cells to secrete new cartilage, the perfect fit between existing and new tissue means the result should be as durable as the real thing. (Since the progenitor cells are the patient's own, in a sense it will be the real thing.) In a beautiful irony, the exact characteristic of cartilage that makes it so reluctant to regrow on its own — its lack of vascularization — is what makes it easier than most other types of tissues to print. D'Lima made his prototype using a 1990-era HP Deskjet 500 inkjet printer. He says it will be several years before his technology is ready to jump from the lab to the operating room. But when it does, it could have a huge impact. Recall what Brian Kelly said about the hip surgeries he does: the limiting factor on who's a good candidate is the health of their cartilage. Once healthy new cartilage can simply be printed into the joint, total hip replacements could be a thing of the past.

For some athletes, a few years is too long to wait for miracle cures. The hype around stem cells — immature cells that have the ability to reproduce themselves indefinitely and mature into multiple tissue types — has outpaced the reality for many years, with only a few treatments using them approved by the FDA for humans in the U.S., none of them having anything to do with sports injuries. Yet athletes like Peyton Manning, Cristiano Ronaldo, Rafael Nadal, and Bartolo Colon are all reported to have received mysterious stem cell injections to make their injuries heal faster. Nadal got stem cells in 2013 in his knee, and then again the following year in his back. His doctor, Angel Ruíz-Cotorro, said the second injection was intended to produce new cartilage in his spine.

It may be that the doctors doing these injections know more than the slow-footed FDA is willing to allow. But stem cell treatments, unlike PRP, are far from risk-free. Jeanne Loring, a biologist who works with D'Lima at the Scripps Research Institute and directs its Center for Regenerative Medicine, says most unregulated stem cell treatments are benign, but some have had severe repercussions, including some deaths. A woman in Los Angeles who got liposuction at a clinic in Beverly Hills opted to have her leftover fat processed into stem cells and injected into her eyelids. Instead of rejuvenating her face, the cells differentiated into bone; when she blinked, according to her doctor, it sounded like castanets.

After a stroke left him with limited use of his left arm and leg, Jim Gass, a former executive for the lightbulb maker Sylvania, was searching for treatment options when he read a story about the pro golfer John Brodie, who attributed his miraculous recovery from a stroke of his own to stem cell therapy. Gass spent nearly $300,000 traveling to Mexico, China, and Argentina for stem cell injections in his spine. Shortly thereafter, he developed a massive, aggressive tumor in his spinal canal. When a surgeon biopsied it, it turned out the cells weren't Gass's. He'd been injected with cells extracted from someone else entirely.

"Everyone here should feel proud to be part of the biggest revolution in the history of humanity. The people who are in this room are at the cutting edge of a movement that's going to take over the world."

The people in this auditorium certainly don't look like they're about to take over the world. Of the 300 or so listening to these opening remarks, a majority seem to be late-middle-aged white folks in what you might call comfort-fit academic wear — rumpled khakis, boxy pantsuits, vests with a lot of pockets. But anyone familiar with the speaker's history knows to expect a bit of hyperbole.

I'm at the SENS Research Foundation's Rejuvenation Biotech Conference. It's taking place at the Buck Institute for Research on Aging in Novato, California, a half-hour drive north of San Francisco. SENS stands for Strategies for Engineered Negligible Senescence. Its founder, Aubrey de Grey, is the speaker onstage now. An Englishman in his 50s, de Grey is part scientist, part philosopher, part provocateur, part media gadfly. He cuts a striking figure, tall and thin with a long ponytail and an equally long chestnut beard. For the past decade, he's been advocating, charismatically if not always convincingly, for the idea that aging is a problem with a solution — that humans can figure out how to live forever, if they throw enough science at it.

Here in Greater Silicon Valley, land of biohackers and health nuts, that's a message with a receptive audience. Half the tech industry's billionaires seem to be using their fortunes to seek an escape from the finality of death. Sergey Brin and Larry Page, the cofounders of Google, launched their company Calico (for California Life Company) with the mission to "harness advanced technologies to increase our understanding of the biology that controls life span." Oracle founder Larry Ellison has channeled a half-billion dollars into antiaging research. Max Levchin, the cofounder of PayPal and Affirm, does 90 minutes of high-intensity cycling every day to stay healthy, but he told me he's not ultimately concerned about the fate of his body: Levchin is a proponent of "uploading," the idea that we'll be able to render our consciousness in the form of data that can then be transferred from one computer to another, ad infinitum.

"Aging is not actually a phenomenon of biology at all. It's a phenomenon of physics," de Grey declares from the podium. "It's something that's going to happen to any machine that has moving parts. Aging is going to consist of the creation of damage because damage is something machines generate as a consequence or a side effect of operation." The radical idea de Grey is trying to impart is that there's nothing radical about thinking we might be able to extend the human life span indefinitely. It's just a matter of identifying all the changes that occur in an aging organism and finding ways to reverse them, one at a time. "Preventive maintenance works in actual real life with inanimate machines," he says. "We have vintage cars that are ten times as old as they were designed to last, and we have them because of preventative maintenance."

De Gray's wide-open optimism sits uneasily alongside the cautious realism of the working researchers who take the podium. First up is Pinchas Cohen, dean of the University of Southern California's School of Gerontology. Cohen's talk is about mitochondrial peptides, protein building blocks manufactured inside the cell's power plant. He zeroes in on two peptides of special promise. One called humanin improves blood sugar levels and appears to confer protection against diabetes and Alzheimer's. People with a centenarian in their family tend to have higher levels of humanin. "We believe humanin injections act as a caloric restriction mimetic therapy," Cohen says in a strong Brooklyn accent. That is, it activates the same cellular mechanisms triggered by extreme low-calorie diets, which have been shown to extend life span in mice and worms. Another mitochondrial-derived peptide, MOTS-c, works in a different way. In a trial of mice put on a high-fat diet, the ones treated with MOTS-c gained much less weight and didn't suffer the same problems with insulin sensitivity and fatty liver as the control group mice. "We like to think of it as an exercise mimetic rather than a dietary mimetic," Cohen says. At this, you can see people in the audience sit up a little straighter. Then, the letdown: the mice didn't live any longer, and no one knows why. "Let me be first to say the fact that we didn't see an increase in life span was a surprise and a disappointment," Cohen says.

Judith Campisi, a professor of biogerontology at the Buck Institute, gives a talk on cellular senescence, the process whereby cells stop dividing and signal to other nearby cells to do the same. First discovered in the 1960s, senescence happens in response to a number of things, chief among them cancer-causing genetic mutations. Over time, as the body's cells divide and undergo genomic replication, more and more of them acquire mutations, resulting in a buildup of senescent cells. When cells senesce, they release a number of organic compounds that produce an inflammatory response, including cytokines and prostaglandin. Chronic inflammation, she notes in an ironic tone, is a known risk factor for development of cancer. "If you're not depressed, you haven't been listening," Campisi says. "You need this response to be cancer-free. But if you live long enough it's going to drive cancer." Campisi concludes with a slide showing the effects of life-extension interventions in a variety of organisms: roundworms, fruit flies, mice, and humans. With each rung up the ladder of biological complexity, the best longevity treatments decrease in their effectiveness. Scientists have figured out how to get roundworms to live ten times' their normal life spans, but the longest-lived humans continue to die around age 115, as they always have. "Maybe evolution is trying to tell us something?" she says.

Taking a break from the talks, I wander the atrium. Over at the snack bar, a man is asking what additives are in the iced tea. A young woman studies poster boards describing experiments. Affixed to her forehead is a device that sends electrical current into her brain. It's supposed to improve mood and focus, she tells me. Then I spot someone I recognize from a photo. His name is Jason Camm, and he's the chief medical officer for Thiel Capital, one of the investment firms run by billionaire Facebook investor Peter Thiel. Small in stature and handsome, he's wearing a soft-shouldered blazer and leather sneakers. I know a little about his background: before going to work for Thiel, he was an osteopath and nutritional therapist in the U.K. with a high-performance clientele of Formula One drivers and Premier League soccer players. I walk over to where he's sitting with his laptop and introduce myself.

"I'm not interested in talking with you," he says abruptly. I don't even have to ask why. It's because of the blood thing, obviously.

A year earlier, I had interviewed Thiel about his investments in a number of biotech companies all working, in one way or another, to solve problems connected with aging and death. Thiel is a famously idiosyncratic and elliptical thinker, but on the question of mortality his position is straightforward: he's against it. In fact, he thinks anyone who claims to be at peace with the idea of dying has simply given in to a powerful psychological defense mechanism rather than live in horror of something we can't do much about. As a person with effectively unlimited financial resources, Thiel doesn't have to accept much of anything he doesn't want to. After talking for a while about the companies he was funding — one sought to use gene therapy to turn white blood cells into factories for therapeutic proteins; another was working on growing replacement bones in the lab — I asked about what longevity treatments Thiel found compelling enough to practice in his own life. Thiel neither answered the question directly nor evaded it entirely. "There are all these things I've looked into doing. I haven't quite, quite, quite started yet," he said. He listed a few and walked me through their pros and cons. Extreme caloric restriction "feels a little too painful," he said. "There are probably a lot of catastrophic approximations that happen. You end up hurting yourself by starving yourself too much." Besides, there's no proof yet it works in humans. The diabetes drug metformin has attracted a lot of interest from longevity researchers. "It has a side effect of suppressing some of the sugar pathways that tend to get used disproportionately by cancer cells," he noted. "I'm looking into it. Not yet."

I had read that Thiel was a taker of human growth hormone, so I asked him about that. "Well, I've looked into the HGH stuff. It's probably also very underexplored," he said. "There's always a worry that it increases the risk of getting cancer but then it probably has these benefits of making it less likely that your muscle mass wastes, and you're less prone to a hip accident or something like that. I suspect we're just a little too biased against all these things in society," he said.

"I'm not convinced yet we've found a single panacea that works," he added, which I took as a conclusion of the discussion. But then he went on. "It's possible there exist single-point things that could work. I'm looking into the parabiosis stuff, which I think is really interesting." By "parabiosis," I knew, he was referring to experiments showing that fusing the circulatory systems of two mice — one young, one old — so that they shared a blood supply somehow caused the old mice to outlive their life expectancy by a considerable margin. Not only that; the old mice became, in appearance and behavior, young again. Parabiosis is something of an obsession in the live-forever set. Covering Silicon Valley, I'd heard various rumors about one tech billionaire or another who supposedly was paying big bucks for transfusions of young blood. "It's one of these very odd things where people had done these studies in the nineteen fifties and then it got dropped altogether," Thiel said. "I think there are a lot of these things that have been strangely underexplored."

Rather than pin his hopes on one Fountain of Youth–like anti-aging cure, he suggested he shared Aubrey de Grey's view of the body as a machine that can be kept young through constant vigilance — no surprise, seeing as Thiel is a major backer of the SENS Foundation. "Think of your body as having tons and tons of little policemen and little criminals," he said. "The policemen are very good at catching the criminals, but if a criminal goes uncaught for a day, you die. It's possible there are a few core mechanisms that drive [aging] and if we could discover them, we could push the envelope a lot."

Jump ahead a year. Just a few weeks before the SENS conference, I heard about a biomedical startup called Ambrosia running a clinical trial of a new therapy consisting of transfusions of young blood plasma from donors under 25 into patients over 35. The founder, Jesse Karmazin, was a Stanford-trained physician who told me he had been "really impressed" by the data from animal studies of parabiosis as well as human studies conducted in China, Russia, and India. "Everything that's been studied seems to improve," he said. "It seems to span the brain, heart, kidneys, muscles. It's almost like there's a resetting of gene expression."

Karmazin hadn't thought too much about the potential market for his product because he assumed it would appeal to basically anyone who could afford it. "I think there's going to be a great demand for this kind of treatment, assuming it works." A former elite athlete with a prosthetic leg — he rowed for Team USA in the Paralympics — he predicted that parabiosis would soon be among the many weird-sounding medical technologies athletes use to maintain their bodies. "It's shown to improve recovery after injuries. It seems to increase energy and muscle strength. I would imagine there are a lot of athletes who could potentially benefit," he said.

Unusually for a clinical trial, Ambrosia's study was "patient-funded"; people on the receiving end of the plasma paid $8,000 each to participate. Karmazin told me he didn't have any investors yet — although, he added, he had been approached "sort of surprisingly early" by a representative of one Silicon Valley investment firm. Surprising, he said, because he hadn't advertised the existence of his company or the trial, except to find participants. But he certainly hadn't been looking to raise money. The person who approached him, he said, was a guy by the name of Jason Camm, from Thiel Capital.

Hmmm.

The story I published stuck pretty close to the facts: Peter Thiel had expressed strong interest in parabiosis; a person who worked for Thiel had expressed interest in a company offering parabiosis; a spokesman for Thiel denied that his boss practiced parabiosis. That was mostly it. But in the time in between my conversation with Thiel and my conversation with Karmazin, the former had gone from Silicon-Valley-famous to nationally notorious. First, he had been revealed as the secret financial backer behind a $140 million invasion-of-privacy lawsuit by the wrestler Hulk Hogan that effectively put an entire publishing company, Gawker Media, out of business. Then he had endorsed Donald Trump for president, making him the only prominent figure in Silicon Valley to do so, and further stuck his neck out by speaking for him at the Republican National Convention and donating more than $1 million to his campaign.

So when my story came out, there was, shall we say, a certain level of interest in all things Peter Thiel, especially anything that made him seem like a James Bond villain. Matt Drudge put it at the top of the *Drudge Report*. "Is Peter Thiel a Vampire?" asked the *New Republic* in a headline. "Peter Thiel Wants to Inject Himself with Young People's Blood" declared the comparatively low-key *Vanity Fair*. The author of *Interview with the Vampire*, Anne Rice, weighed in: "Blood is something that everyone has to give. I don't see why it has to only involve rich people." It even inspired an entire episode of HBO's *Silicon Valley*, in which the billionaire investor Gavin Belson shows up to a meeting with his "blood boy," a healthy teenage donor, and proceeds to receive a transfusion.

"Out of all the crazy things in this campaign, the vampire accusations were the craziest," Thiel later told *New York Times* columnist Maureen Dowd. He hasn't talked to me since.

And neither, apparently, was Jason Camm going to — at least not on the record. I can't say I blamed him. I also can't say I minded terribly. I may be a little obsessed with what science has to say about staying healthy and vital as I get older, but I could give a fig about living forever. Inasmuch as death often issues from diseases that cause a great deal of suffering — cancer, diabetes, Alzheimer's — the scientists at this conference were working with the noblest of aims. But the people who had come here in hopes of finding out how they could escape the fate of all their ancestors — that's a different matter. The very idea strikes me as self-centered, immature, emotionally stunted — as do, frankly, many of the people who seem most intent on it. Maybe that's a defense mechanism, as Peter Thiel says. If so, it's a good one.

Any able-bodied person who gets giddy at the thought of a drug that eliminates the need to exercise has things backward, as far as I'm concerned. You might as well enter a marathon and then pay someone else to run it for you. The point of living isn't to survive any more than the point of playing sports is winning. The idea that science and medicine can repair damage to our bodies, rejuvenate our joints and muscles, turn back the hands of time — that's very exciting to me. And the fact that exercise is the most potent antiaging treatment ever discovered feels like poetic justice. It's beautiful precisely because exercise — whether it's ecstatic midnight dancing or sweaty 6:00 a.m. suffering — is already so manifestly its own reward. If there's a good reason to desire eternal youth, it's never having to go without that feeling.

Youth, to me, isn't about distance from death. It's about the opportunity to challenge yourself, to grow, to strive without limitations, to feel like a beginner because there's so much room for improvement and so much joyous work and play ahead. The athletes who have inspired me as I've gotten older — Meb Keflezighi, Carli Lloyd, Roger Federer, Jaromir Jagr, Hilary Stellingwerff and Catharine Pendrel, Brett Favre and Donald Driver, Kerri Walsh Jennings, Alex Martins — are all people who have won a lot in their lives, sure, but that's not why they inspire. Like them, I just want to put my whole heart into living and playing while I can. That's all anyone could ask for. That's a lot.

EPILOGUE

Works for Me

When I told people I was working on a book about all the different things athletes do to stay healthy and competitive as they age, I was often asked: Have you learned anything so compelling that it changed what you do in your own life? It's a good question.

I'm not an elite athlete. I'm not even a particularly impressive recreational athlete. I'm just a guy on the precipice of middle age who enjoys cycling, running, soccer, and tennis and hates it when my body holds me back from doing as much as I have time for. If I can push myself to my physical limit a couple of times a week and still have energy left over to crawl around on the ground with my daughter, I'm satisfied.

In a way, I'd argue that that makes me a useful filter. Top pros have effectively unlimited budgets and no demands on their waking hours except to make themselves fitter and more skilled. As we've seen, elite athletes will do almost anything to add a couple of years to their careers, even interventions that lack scientific support or damage their health later in life. If you want to spend $30,000 to sleep in an oxygen pod wearing infrared pajamas, knock yourself out, I guess. If you're only interested in stuff that works, is safe, and is doable on the schedule and budget of a working professional, trust the guy with a day job, a bad back, and a baby.

In that spirit, here are a few of the things I've added to my own performance routine, such as it is, after learning about them from writing this book.

Periodize, periodize, periodize. If there's one concept whose centrality has been drilled into my head over and over, by the likes of Raymond Verheijen, James Galanis, and especially Trent Stellingwerff, it's the importance of periodizing for performance and the risks of not doing so. For an elite athlete, that usually means a highly structured training program whose volumes build up and taper off in a planned way to yield peak fitness at the required time. For me, it's not so much the structure but the principles I try to adhere to: ramping up training gradually, preparing my body specifically for the demands I plan to place on it, and avoiding the buildup of fatigue or the sorts of sudden jumps in volume or intensity that lead to avoidable injuries. If you invite me to play soccer and I haven't been keeping in soccer shape, or if I'm nursing an injury I could play through, I say no. Benching yourself sucks, but not as much as getting hurt and missing an entire season sucks.

Making unloading and mobility part of every workout. A key difference between elites and the rest of us is how you define when a workout is over. Strength coaches and athletic trainers like to talk about "loading" and "unloading," the idea being the former should always be followed by the latter in some reasonable proportion. I now think in those terms. Unloading can mean yoga, foam rolling, ice tubbing, aqua running, meditating, or just taking a nap. It encompasses recovery as well as range-of-motion work that prevents the sort of movement limitations and compensations that can build up over time and lead to injuries. For me, it has meant going from someone who considered happy hour a valid cool-down routine to becoming a fanatic about stretching and self-massage, with a closet full of straps, bands, foam rollers, PVC pipes, and lacrosse balls to show for it. A commitment to unloading and mobility is tough for people with busy lives; if you only have a few hours a week to devote to fitness, the temptation is to extend your run half an hour and skip the stretching. If you're injury-prone like me, that's a bad trade-off.

Polarize it. This is another one I incorporated after hearing about it from Trent and Hilary Stellingwerff. The idea of polarization, you'll recall, is that only 20 percent or less of your workouts should be at high intensity, and the balance should be performed at such low intensity that they require little or no recovery. Again, I don't stick to any formal program of polarization, but I do try to avoid what Trent says is the most common mistake athletes — including elite ones — make in their training: going too hard on easy days, and then not being able to go as hard as you want to next time out because you're still fatigued from a workout that didn't serve any particular purpose. On the flip side, I make my hard workouts both shorter and more intense than I used to. Studies of elite older athletes have shown that one of the ways they stay competitive as they age is by getting more deliberate in their training, focusing their limited time on the skills practice or fitness work that is most difficult for them. For me, this often means taking two minutes before I start to write up a plan on a Post-it. A little intentionality goes a long way.

Eating for muscle. In case I didn't make it clear enough earlier, a great deal of the nutritional "science" being peddled to athletes is bunk. If you're eating a generally healthy diet — lots of veggies and whole grains, not too much sugar or processed stuff — you're probably fine. But if you want to avoid losing muscle as you age, it's worth making a couple of tweaks. On the advice of Asker Jeuk-endrup, I've upped the amount of protein in my diet and the number of times I consume it during the day (although I stop short of having a protein shake right before bed). A side benefit of this: according to Chris Jordan, the director of exercise physiology for the Johnson & Johnson Human Performance Institute, adding protein to anything you eat has the effect of lowering its glycemic index. So if I want to have an oatmeal-chocolate-chip cookie without feeling a sugar crash, I'll put a smear of almond butter on it. I also try to consume three to five grams of creatine powder daily, usually in a smoothie or a glass of milk, right before or after a workout. The science behind creatine is hard to argue with, and it's had a noticeable effect on my ability to build and maintain muscle.

Collagen/gelatin/bone broth. Yes, it's a huge fad, but it's the rare nutrition fad with data to back it up. One nice thing about bone broth being trendy is it's easy now to find prepackaged varieties that don't taste gross. On the other hand, making it yourself from a chicken or turkey carcass is the simplest thing in the world and weirdly satisfying.

Midfoot running. This one's not for everyone, and I'm going out on a limb a little bit in advocating it. The great barefoot-running craze of the mid-2000s crested and crashed long ago. Nowadays you see lots of older runners pounding pavement in "maximalist" shoes made by Hoka or Altra, with soles as thick as a *Game of Thrones* paperback, and not worrying about whether they're landing on their heels or toes. I'm not hoping to spark a comeback of those ridiculous finger-shoes. But after wading through conflicting research and talking to a few biomechanists, I've becoming convinced there are some concrete benefits for runners to forgoing a little cushioning and learning to strike the ground with your midfoot or forefoot rather than your heel.

Your legs are springs. The stiffer the spring, the more efficiently the forces you put into the ground return to your body, propelling you forward. Midfoot running requires a high degree of stiffness from the foot and ankle, putting the onus on the lower leg muscles to maintain that stiffness. As people age into their 50s and 60s, their running styles tend to change in characteristic ways. As their tendons lose their stiffness and their muscles lose strength, they compensate by relying more on their hips to absorb and generate the forces involved. Running researchers call this the "distal-to-proximal shift of joint powers." In a sense, you could say a proper midfoot gait is a "young" running pattern. All things being equal, if you're running to keep your body young, doesn't it make sense to run like it?

There are a couple of caveats here. The first is that runners, even expert ones, are notoriously bad at identifying their own strike patterns. If you're going to try to change your strike, I suggest working with a coach or seeking out a treadmill with sensors that can show you how you're landing. The other caveat is that midfoot striking may make you slower over long distances. Studies have suggested the most energy-efficient running style for most people is the one that comes naturally to them. Jay Dicharry, a biomechanist who's generally a fan of minimalist footwear and the associated running style, readily concedes this. "Most folks are spending more energy to run with that style because you're using more joints," he says. But crucially, Dicharry also points out that forces absorbed by your calves are forces that don't have to be absorbed by your knees, hips, or lower back. "You're robbing Peter to pay Paul," he says.

Worse times in exchange for less pounding on my spine is a trade-off I'm happy to take.

Harder, not heavier. For athletes with a history or injury or physical limitations — and past a certain point, that's all of us — the key to optimal fitness is finding ways to separate desirable training stresses from undesirable ones. If you have access to an AlterG treadmill or Kaatsu bands, that's great. If you don't, there are still plenty of ways to embrace this concept. Instead of adding extra weight to an exercise, I'll add a balance element, like doing push-ups with my hands on medicine balls, or a second force vector, like a resistance band around my knees while squatting. Focusing on smaller, neglected muscle groups isn't a recipe for getting huge but it's great for functional strength and avoiding injuries.

Self-talk. Before I learned what a powerful performance tool it can be, I never gave much thought to the particulars of my internal monologue. Now I do. I've discovered that I perform best when challenging and exhorting myself rather than encouraging, praising, or criticizing. Self-talk is particularly handy in lonelier pursuits like running and cycling. I've also borrowed Meb Keflezighi's trick of going into events with tiered goals. I use the Strava app to record all my bike rides. On days when I plan to go hard, I always have multiple goals for the climb I'm attacking: setting a new PR, top 10 for the day, top 5 percent for the year, et cetera. The fine-grained competition with myself and others on Strava inspires me to ride much, much harder than I would otherwise, even on days where I can tell I'm not my fastest.

Starting over. This book began with what was, for me, a new beginning. On the wrong side of 30, I took up soccer, a sport that made me feel clumsy, slow, and overwhelmed. It was wonderful. My conversion to the religion of cycling came after tweaking my back yet again on the soccer field and feeling like I needed to try a less-punishing sport. Now I spend my weekends gasping for breath and cursing my legs, wondering how many more turns there can possibly be before I reach the top of the hill. It's wonderful.

Challenging your body in the same ways day after day for decades on end is an efficient way to chew it up. Challenging it in different ways is the perfect cure. But there's more than that to be said for starting a new sport. I love to marvel at the abilities of elite athletes and, like most nonelites, I envy what they can do. But they should envy us too. Greatness is a burden, and there's a profound freedom in knowing you'll never have it. There's nothing like trying something new and sucking at it, then sucking a little less every day you keep at it. Until science makes it possible to get younger, getting fitter, faster, and better at what you love will remain the closest thing most of us have.

ACKNOWLEDGMENTS

Writing this book was a marathon, not a sprint. An ultramarathon, maybe. It took me more than four years to see it through from idea to publication, a period during which I moved across the country, got married, changed jobs, moved again, and had a baby. Spend four years working on a nonfiction book and you end up with an awful lot of people to thank.

My agent, Daniel Greenberg, was the perfect blend of patience and persistence. The first time I talked to him — at the suggestion of Colin Dickerman, thank you very much — I was lying flat on my couch in Brooklyn, waiting for my spine to heal up enough that I could start rehab. (Shout out to Dr. Eric Elowitz of New York Hospital and Tony D'Angelo of Professional PT for getting me back on my feet.) Throughout this process, Daniel was a source of encouragement, a voice of calm, and a keen dispenser of editorial advice. His associate Tim Wojcik deserves his own round of applause as well.

On the subject of patient editing, my editor, Susan Canavan, gave me an awful lot of leash, including a pair of deadline extensions after the original deadline turned out to be my daughter's due date. I hope I've repaid her trust. In addition to Susan's oversight, the finished product also benefited from the attentions of Rebecca Springer and Melissa Dobson, while Lisa Glover kept the whole enterprise together. Megan Wilson was my PR ace. My research assistant, Sasha Lekach, did a bang-up job rounding up answers to my often abstruse questions.

Four years sounds like a lot, but it could easily have been six or eight had it not been for my supportive and wonderful bosses at *Inc.* magazine, including Eric Schurenberg, Jim Ledbetter, Jon Fine, Laura Lorber, and Kris Frieswick. Rather than view my book project as a competing claim on my time, they welcomed it as a source of great business stories. It was for *Inc.* that I first wrote about the business of injury prevention, cryotherapy, recovery, and numerous other fitness- and performance-related topics. As accomplished authors themselves, Jim and Jon also had plenty of valuable advice, more of which I should have followed.

A great number of people shared their time and expertise to help me understand what's happening in the elite precincts of sports and sports science. No one's generosity and insight went farther than that of Trent and Hilary Stellingwerff. I'm not sure how I would have written this book without the time I spent in gorgeous Victoria, where the Stellingwerffs opened up their home and workplaces to me, fed me, introduced me to the local athletic community, and even assigned me homework reading. Even by Canadian standards, they are extraordinarily nice people.

It should be obvious from reading this book that I also owe great thanks to Meb Keflezighi and Bob Larsen, Raymond Verheijen, Marcus Elliott, Phil Wagner, Tom Incledon, Daniel Chao, James Galanis, Catharine Pendrel, Alex Martins, Mackie Shilstone, Kirsty Coventry, Steven Munatones, Nirav Pandya, and Tony Ambler-Wright for their time and openness. Less visible was the assistance of Rob Gathercole, Anthony Katz, Ricardo Geromel, Morgan Oliveira, Yelena Gitlin Nesbit, Michael Cooper, Kate Hyatt, Dan Bigman, Eleanor Prezant, Sara Edwards, and Rachel Sattler, all of whom pitched in behind the scenes with ideas, introductions, or other assistance and support. I don't have space to thank everyone I interviewed, but they all have my gratitude.

One of the best things about being a journalist and writer is spending time around other journalists and writers. They're the best people. I can never thank Jonah Weiner enough for talking me into taking up cycling, or for listening to me puzzle out various conundrums of structure and theme on our long rides through the hills of Berkeley and Oakland. Tagging along on one of those rides, Andrew Vontz suggested the title for this book. Matt Haber is my go-to sounding board for any stupid idea that pops into my head, and he also officiates a hell of a wedding.

No acknowledgments are complete without the moms. From the time I was young, Judy Bercovici made it clear that nothing I could do with my life would make her prouder than my being a writer. I truly don't know what she was thinking, but I'm grateful. Susan Mostel was an avid and tireless unpaid research librarian. Quite a lot of the information she turned up made its way into the finished product.

Then there's the mom who wasn't a mom yet when I started this journey. Aly Mostel supported this undertaking in every conceivable way from the day I arrived home at our apartment in Brooklyn and told her I had a book idea. Since she worked on sports and wellness books for a living, when she told me it was a worthwhile idea, I believed her. Since then, she has been my indispensable advisor, publicist, wrangler, go-between, and all-around moral supporter. She took care of me in the months it took me to recover from my back injury, moved across the country when I got a job in San Francisco, made me the happiest dad in California, and then did the lion's share of parenting for the next five months while I finished this book. Aly, thank you for everything. And Ramona, thank you for being my favorite inspiration to be stronger next year. Too soon, you'll be able to beat me at everything, and I can't wait.

Lastly, thanks to Ciaran Gorman, Julie Gerstein, and the other members of Gotham United. Those Friday nights changed my life.

A NOTE ON SOURCES

In researching the sports science around athletic longevity, a number of books and other resources were especially helpful to me. Bruce Grierson's *What Makes Olga Run?: The Mystery of the 90-Something Track Star and What She Can Teach Us About Living Longer, Happier Lives* (Henry Holt & Co., 2014) is full of information on the science of superagers and masters athletics. Bill Gifford's *Spring Chicken: Stay Young Forever (or Die Trying)* (Grand Central Publishing, 2015) is a hilarious tour through the world of longevity science and pseudoscience. David Epstein's *The Sports Gene: Inside the Science of Extraordinary Athletic Performance* (Current, 2014) is an authoritative examination of how genetics, biology, environment, and training interact to produce world-class athletes. *Running: The Athlete Within*, by David Costill and Scott Trappe (Cooper Publishing Group, 2002), is a foundational text in the field. Mark McClusky's *Faster, Higher, Stronger: The New Science of Creating Superathletes, and How You Can Train Like Them* (Avery, 2014) offers a highly readable overview of contemporary sports science. For a book on periodization, *How Simple Can It Be? Unique Lessons in Professional Football: Behind the Scenes with*

Raymond Verheijen (World Football Academy, 2015), by Frank van Kolfschooten, is absurdly entertaining.

For anyone trying to keep up with the latest findings around fitness and injury prevention, Gretchen Reynolds's *New York Times* columns and Alex Hutchinson's *Sweat Science* blogs are must-reads.

I've included several citations from *Open: An Autobiography*, by Andre Agassi. It deserves its reputation as one of the best sports memoirs ever written. Thanks to J. R. Moehringer both for writing it and for giving me his blessing to quote generously from it. My writing about tennis was also informed by *Strokes of Genius: Federer, Nadal, and the Greatest Match Ever Played*, by L. Jon Wertheim (Houghton Mifflin Harcourt, 2009), and *Late to the Ball: Age. Learn. Fight. Love. Play Tennis. Win.*, by Gerald Marzorati (Scribner, 2016).

Several of the quotes in chapter 7 are from a blog post Catharine Pendrel wrote after her bronze-medal-winning race at the Rio Olympics. For readability's sake, with Pendrel's permission I presented the blog-post quotes in the same format as quotes from our in-person interview. I also drew from that post for my account of the race. You can read the whole thing at http://cpendrel.blogspot.com/2016/08/olympics-can-be-magic.html.

INDEX